# PAWS & Listen

## TO THE VOICES OF THE ANIMALS

I would like to dedicate this book to my parents:
my mother, Lenore, and my late father, Bill Wille.

A special dedication goes to Smurfie, my little angel dog,
and all the animals, past and present, who have touched my life
and made this book possible.

*Jenny*
## SHONE

**Fourth Edition Published by**

Jenny Shone

PO Box 464, Walkerville, 1876

E-mail: jenny@animalhealing.co.za

www.animalhealing.co.za

Copyright © 2021 Jenny Shone

First edition published by Jenny Shone, 2005

All rights reserved under international copyright conventions. No part of this book may be reproduced, stored in a retrieval system, or transmitted in any form or by any means electronic, mechanical, photocopying, recorded or otherwise without written permission from the publisher.

Whilst every care has been taken to check the accuracy of the information in this book, the publisher cannot be held responsible for any errors, omissions or originality.

ISBN: 978-0-620-97302-1(print)

ISBN: 978-0-620-97303-8(e-book)

# Contents

Author's Note ...................................................................... 6

Acknowledgements ............................................................ 7

About the Author ................................................................ 9

Foreword .......................................................................... 13

**Chapter 1**
Smurfie's story ................................................................. 15

**Chapter 2**
Conversations with pets ................................................... 21

**Chapter 3**
The extraordinary is ordinary - X-ray vision and body scan .......... 53

**Chapter 4**
The intuitive experience - tracking through Gestalt ................. 79

**Chapter 5**
The contact experience during grief - heavenly voices ............. 95

**Chapter 6**
The soul experience - death and rebirth ................................. 111

*Photographs*
The animals behind the voices .................................................. 119

*A Special Acknowledgement to*
*Amelia Kinkade*

The title
"Paws & Listen"
was inspired by her book
"Straight from the Horses Mouth".

She has been my greatest inspiration
in all my communication work with animals.

*A special Thank You!*

# CONTENTS CONTINUED

**Chapter 7**
Talk to the wild side - and voices from the ocean .................. 127

**Chapter 8**
The healing experience .......................................................... 157

**Chapter 9**
Horse sense - the horses have something to say ..................... 163

**Chapter 10**
Smurfie's rebirth and gift ....................................................... 175

Meditations and Exercises ..................................................... 179
The Lion Heart Project .......................................................... 193
Important information: Protect your cats and dogs
    from harmful foods and substances ................................... 196
Animal Organisations ............................................................ 200
Articles written about and by Jenny Shone ........................... 207
Recommended reading .......................................................... 208
Index ...................................................................................... 210

# AUTHOR'S Note

*My angel dog, Smurfie*

I decided to write this book to share some of the most memorable experiences I've had with the animals that have crossed my path over the years. My greatest desire is to improve the lives of all animals and our relationships with them.

Everyone has the ability to achieve their greatest desires, no matter what they are – just as I have done.

I look forward to sharing with you some of the amazing stories the animals have shared with me.

Enjoy!

**Jenny Shone**
Johannesburg, October 2005

# Acknowledgements

Thank you, Smurfie, for touching my life in such a big way and helping me to share your experience with others.

I would like to thank all the animals whose stories appear in this book and also their humans for agreeing to let me print their stories.

Thanks to my husband, Alan, for his constant patience and support over the years. To my sister, Sandra Parsons for being a great older sister, and to my mother, both of whom have given me endless support from the very beginning and who continue to support me in everything I do. To my sister, Linda Brown, for having confidence in me, for encouraging me to start this book and for all the hours spent helping me to write it.

To Alba Delport, who took on the task of producing this book, and has given me endless advice. Thank you to Judy Geyer for all her help in designing the cover and everything else she has done for me.

Thank you to Dianne Kruger for her write-up on the history of the Alpacas. I would also like to thank Gail Kleinschmidt for her generosity in providing the information on the Lion Heart Project.

A special thanks goes to the following for their constant support: Kumi, my spiritual adviser, for helping to keep me focused and grounded; Sandy Carter, my partner in The Animal Healing Centre for all her enthusiasm, help and support; Sandra and Mike Parsons, my sister and brother in-law in the UK; and my mother-in-law, Meg Brown, who has given me lots of advice on my writing techniques.

I thank all the animals I share my home with, for bringing me back down to earth from time to time with their humor and for reminding me that they are the ones writing this book.

*Jenny and Babanjala, an nyala*

# ABOUT THE Author

My sister, Jenny, has been an inspiration all my life. Our parents still have the tape recording of her first words, at the age of two, which show the beginning of her animal connection. They were: "I want an elephant!" (She got a hamster a few years later).

I remember a day when our family went to visit a friend, Brian, who was an expert on snakes and collected them as a hobby. Jenny and Brian later disappeared upstairs. A few moments later, Brian appeared with a 16-foot python draped around his neck, closely followed by Jenny with a 14-foot python draped around her neck. They ran down the stairs and, with the pythons, plunged into the swimming pool, where they swam together happily.

Many years ago she wrote her first book (although never published) about the life experiences of our father during the Second World War.

Jenny has always had an all-consuming passion for animals and all other interests have taken second place. At the age of sixteen, she went into partnership with a good friend, Wendy Brode, and opened her first horse riding school.

Later, she specialized in teaching disabled children. At the age of eighteen, she joined and worked with the world famous Lipizzaner horses in Kyalami and even visited the Lipizzaner Spanish riding school in Vienna.

She later moved to Natal, taking all her horses and riding school with her, and it was here that she married Alan. She took up modern dancing and later qualified as an international aerobics

# ABOUT THE AUTHOR

instructress, taking part in a nine-hour aerobic marathon. She also traveled to various aerobic seminars around the country.

When Jenny and Alan, together with her riding school, moved back to the Johannesburg region, she took an interest in the police and became a police reservist. Jenny soon went up in the ranks and became an Inspector.

However, after twelve years had passed, her police duties left no time for her work with animals and she resigned.

Coming from a music-loving family, Jenny has always maintained her interest in music and is busy pursuing her love of singing. She is often to be seen singing to animals in zoos and game parks. She also spends time taking her dogs to "school" where they have won prizes for their agility abilities. Jenny has studied Tai Bou and Martial Arts and is currently studying the art of Tai Chi.

Jenny had the good fortune to meet her little dog, Smurfie, whose passing cemented her determination to enquire about life and spirituality. Her determination to find an answer has led her on an amazing spiritual journey and her introduction to Kumi, a master in metaphysics, who runs The Mystical Centre in Park Town, Johannesburg. Kumi runs courses in healing, psychic development, meditation and martial arts, and gives talks on all aspects of spirituality. He would later become her spiritual advisor and friend and encourage her to continue her journey. This has resulted in Jenny's deep understanding of her connection with animals and her realization that she has been communicating with them all her life.

Her book now gives us the opportunity to understand and experience this wonderful gift ourselves.

Jenny comes from an artistic family and has a love of painting.

# ABOUT THE AUTHOR

Recently, she became involved with painting portraits of animals that have crossed over. Her portraits capture their essence.

Jenny has been called in and asked to consult for a script writer of a locally-produced television drama series set in a veterinarian hospital due to be screened early in 2006. Recently she was also asked to take part in a wild life documentary that will include telepathic communication.

I feel that her work with animals is going to take us all on a wonderful journey starting with this book. I look forward to your future projects, Jenny.

With love,

Linda

**Ian Melass**
*Wild life expert and manager of the Lanseria Lion Park,
Johannesburg, South Africa.*

# Foreword

"Paws and Listen to the Voices of the Animals", is an animated, interesting and informative introduction for every person and family who carry the responsibility of having pets at home or who work with animals. This delightful, humorous, non-scientific book will be a good place to start.

Jenny's all consuming passion for animals is evident throughout the book and as she recounts her numerous adventures with a broad selection of animals domestic and otherwise, one begins to realise how lacking we often are in truly understanding their varied behavioral patterns and the life lessons, love and loyalty that they give so generously.

Smurfie's story, conversations with animals, intuitive experiences, times of grief and joy, death, the healing process, rebirth all add to the richness of the varied animal experiences.

The growing moral concern for animals and their welfare by the general public, is putting pressure on animal behaviorists and scientists to investigate animal consciousness, suffering and their ways of communicating with us.

This book is educative in an accessible, easy to read format and will be enjoyed by anyone who has an interest in these noble creatures.

Enjoy the read!

*Jenny and a white lioness connecting at the Lanseria Lion Park*

# CHAPTER One

## SMURFIE'S STORY

*To all the readers of this book…*
*Look deep within your hearts,*
*find and feel the love,*
*go forward and spread this love to others.*
*For it is this Love that will heal our planet…*
*Know that this is a message*
*that comes straight from my heart.*

**Daisy**

# SMURFIE'S STORY

Daisy was the first Jack Russell I had the privilege of sharing my home with, having always shared with big dogs in the past. When she was two years old, I decided to let her have a litter of puppies. She obliged and had two of the most beautiful Jack Russell puppies, one I named Snoopy and the other Smurfie. I loved them both dearly. Make no mistake, I loved all my animal companions, but Smurfie and I developed a very special bond. She was indeed my soul mate in doggy form. We had lots of fun together with Smurfie always at my side, keeping me company, protecting me, being there for me when I was down, and we grew closer and closer.

One day I noticed a lump on Smurfie's leg, so off we went to the vet. They decided to remove the lump and, in doing so, discovered it was a cancerous growth. I was devastated. But Smurfie came out fighting and made a full recovery. We played ball, went swimming (her favorite activity) and went for long walks.

Two years later, just before an Easter weekend, my gardener was away on leave and my husband was working in Oman. I was alone at home, and Smurfie and I were playing in the garden with the hosepipe when the phone rang. I dropped the hose and went to the phone. It was Alan phoning from Oman, so I left Smurfie playing with the hose and finished my phone call. When I went back outside, Smurfie was in the water playing madly. I turned off the tap and we went inside. Later, I noticed that my little dog was very bloated and thought it must be the water she drank while playing. Later, when she wouldn't eat, I decided to take her to the vet for a check up.

The vet said to me she would be okay and he gave me some tablets to give her. Much later, Smurfie started coughing up bits of blood. I phoned my vet in a panic and he said, "Don't worry, she will get worse before she gets better."

# SMURFIE'S STORY

During the night, Smurfie spent most of the time with me, sleeping on my pillow and getting up every so often. I eventually fell asleep with Smurfie's head on my pillow next to mine. When I awoke much later, she was gone! I sat up in a panic and saw her lying on the floor in the corner of the room. I jumped up and ran to her, but she was unconscious. I picked up her little body and ran to the phone. I managed to find a vet who worked from home, and a neighbor who would take us to this vet. I sat with her and held her while the vet worked on her, but sadly Smurfie passed away at 3am on Easter Friday.

I felt as if the bottom had fallen out of my world. The pain was unbearable, I felt nauseous and just wanted to die. I now had to go home and tell the others. I felt desperate. My little light had gone out. Suddenly nothing made sense. Where before I had given thanks every day for my life, the animals in it, the beautiful sunsets, now, all I could see was gray and I couldn't go on. I couldn't bring myself to go to the shops, see anyone. My parents came from Natal to look after me. I was going downhill fast, and there was nothing I could do, and nothing I wanted to do. My girl was gone and all I wanted was to be with her. The other pets in my life gave me lots of support and love, but I was feeling too numb to notice. How was I going to spend the next hour, let alone the next twenty years, without her?

Eventually, just when I thought I was reaching breaking point, a friend gave me the phone number of a psychic. I had never been to a psychic before, but I was desperate and so made an appointment.

I walked into the psychic's room and she started talking to me. After a while, she suddenly stopped and asked me who the little black and white dog was at my feet, and who, the more I talked, the more she yapped. It was Smurfie, my little girl.

# SMURFIE'S STORY

She told me that Smurfie was going to come back to me in this lifetime. I was thrilled! At last, there was a shimmer of light at the end of the tunnel. But then I started wondering: how was I going to know it was Smurfie? When was she going to come? What if I missed her?

Every dog I saw, I wondered: "Is this Smurfie, is this her?" Desperate for answers, I started reading books to try and develop my own psychic abilities. I did various courses in Reiki, ESP Healing, Pet Reflexology, anything and everything that would help me to understand what had happened, and also to enable me to help other animals and their humans, who were in need.

Many years later I was thrilled to witness the rebirth of my little Smurfie.

All my research paid off and I felt everything was worthwhile. I felt whole again!

# MEDITATION
## 🐾 Connecting to the earth 🐾

All the meditations and exercises in this book are designed to improve your intuition and develop your senses so as to enable you to hear the voices of the animals in your lives.

This is a short meditation to help you connect to the universe and make it easier for you to communicate with animals.

Once you stop seeing yourself as the Human, and realize that even the ants and beetles, and all other insects and animals, are all part of the same universe, and therefore just as important in keeping the universe alive, you will then start to think of yourself as equal and be able to connect and communicate with all living beings.

Find a place out of doors; if you live in a complex then go to a park where you can sit on the lawn out in the open.

Once you are sitting comfortably on the grass,
imagine yourself getting smaller and smaller
until you are the size of a blade of grass.

Look around you and see the grass all around you.
See the ants scurrying about, listen to the beetles in the grass, the frogs in the distance.

The sky is high above you; the breeze is blowing over you. You are a small blade of grass in a magnificent universe.

Become aware that you are connected to this magnificent universe. Become part of this universe.

*continued...*

Listen and identify all the sounds you can hear,
the birds singing, dogs barking, the traffic in the distance, any
sounds you can hear.

Now, slowly,
start getting bigger until
you are back sitting on the grass.

Become aware of your physical body.
Look around you at the beauty of the earth, the trees, the sky and
all the living things that inhabit this beautiful earth.

Lift your arms above your head and stretch,
allowing the energy to move through your entire body, leaving
you with a feeling of exhilaration
and appreciation for the world you live in.

# Chapter Two

## CONVERSATIONS WITH PETS

*Be kind to the animals in your lives,
for they will take you places
And show you how to live
a productive and exciting existence*

*Open your eyes and look deep
into the souls of your animal friends,
You will find many pleasant surprises
in store for you.*

*All you need do is be open . . .*

**Riff-Raff**

# CONVERSATIONS WITH PETS

## 🐾 Tina and Charlie

Many years ago, I shared my home with two very beautiful Bull Terrier/Labrador crosses, Tina and Charlie. After spending time in the garden, I would walk inside to find the two of them fast asleep in the lounge. I would decide to play a game with them, so, without wakening them, I would turn my back on them and think: "I wonder if we should go and play with the balls now." When I turned around, they would both be sitting up with the balls in their mouths. I would think, "Wow, what clever dogs." What I hadn't realized was that I had just sent them a picture of themselves with the balls in their mouths and they were simply responding.

Many people think that when you communicate with an animal you will hear words coming from the animal, but this is not so. Animals don't communicate the way we do. They communicate by sending thoughts, feelings and pictures to each other. If you can learn how to do this and practise it, then you can also communicate in this way. For example, if you walk out into your garden and see your dog digging a big hole in your flowerbed, what do you do? You shout "Don't dig!" Without knowing it, you have just sent a picture to your dog of himself digging a hole, and he thinks "Yaaaaaay, I'm allowed to dig", and so he carries on. It would be much better to say "Let's fetch the ball so we can play." He will then get a picture of himself fetching the ball, as happened with Tina and Charlie.

How many of you have a problem when your visitors arrive and need to lock your dogs out to prevent them from jumping up with muddy paws? Two of the problems you might face here are: firstly, the dogs believe they have done something wrong and are being punished; and secondly, your being scared of the dogs jumping up puts that picture in their minds, and they think it's what is expected of them.

# CONVERSATIONS WITH PETS

Your dogs feel they are part of the family and should be allowed to welcome visitors. Instead of saying, "Don't jump", try saying, "Keep all four paws on the floor", or simply, "Paws on the floor". The dogs will then get a picture of themselves with all four paws on the floor.

One of my dogs, Riff-Raff, told me a long time ago that his job was to welcome visitors into the house. When I started running my workshops, he would get so excited he couldn't control himself. I was worried he would knock someone over, so I decided to keep him outside while everyone arrived, and then let him in as soon as they were all here. Every time Riff-Raff was shut out while people arrived, he would come in when I opened the door, go straight to the guest toilet and lift his leg on the toilet. When I asked him about this, he said he was angry because I had taken his job away from him.

The following workshop, I decided to let him welcome the people in, but said to him that if he was going to welcome everyone, he would have to calm down and keep all paws on the floor.

The very next workshop, as everyone arrived, Riff-Raff bounded up to them, sat in front of them and waited for them to pat him before running off outside to play. He didn't lift his leg on the toilet that day, either!

While this can be very effective, you need to realize that, just like a child, or in some cases even an adult, they might listen once, but you need to keep reinforcing the message. Tell them how proud you are of them for listening, and please carry on welcoming everyone in a calm manner. After all, they do tend to forget after a while.

Telepathic communication is not something new to your pet. What is new is having a human that can speak their language. Don't be disappointed if they don't respond the first time. It takes practice on both parts. And just like us humans, they also don't always feel like talking. You need to respect that. For instance, they might be busy, have nothing to say, just feel like resting or simply have something better to do, as I found out with Snoopy one day.

I was sitting at the dining room table when Snoopy jumped up onto my lap and looked right into my eyes at close range. I looked into his big soft brown eyes and asked "Snoo, do you want to chat?"

He replied, "But of course."

Snoopy has a history of chasing Thomas, my gardener, so I thought now is my chance. I asked him, "Why do you always chase Thomas?"

To which he answered, "It's only a game and someone has to."

With that, there was a noise outside, and Snoopy jumped off my lap and ran outside at high speed, kicking up dust as he sped off.

I said telepathically, "Hey! Where do you think you're going? I'm not finished yet!"

He ran straight back, sat in the doorway, looked up at me and said, "Can't chat now, gotta go and bark…"

##  What is Telepathy?

Telepathy is a mind-to-mind contact.

Have you ever had the experience of knowing that the phone was going to ring before it rings, and also knowing who was on the other end? Have you ever had a premonition that something good or bad was about to happen? How about knowing what someone was going to say seconds before they said it? All this is a form of telepathy.

People often talk about a left brain and a right brain, the left brain being our analytical practical side, which helps us to analyze and make sense of things. Our right brain is our intuitive and psychic side, which is where our intuition and gut feeling comes from.

Every one of us is born with psychic abilities, but as we get older these abilities start to shut down because we fear the unknown or get embarrassed. Over the years, we are made to analyze everything. If it doesn't make sense, then it isn't true. If we can't see it, it doesn't exist. People say to us, "It's not true, it must be your imagination." After a while, our right brain starts to shut down, blocking our intuitive and psychic abilities, just like a muscle we never use. We push aside all our gut feelings and say, "It can't be, I must be imagining it." Many years later, some of us eventually become aware of our abilities and decide to reawaken our right brain. However, it takes numerous exercises and lots of time to reawaken our intuitive abilities.

The three main senses that animals use to communicate are clairvoyance, clairaudience and clairsentience.

### 🐾 Clairvoyance (mind-to-mind)

Clairvoyance is the ability to see pictures, but not with the physical eyes. For example, close your eyes and picture a slice of chocolate cake. It can make your mouth water even though there is no cake there. You have seen a picture without using your physical eyes.

When animals send messages, it is like watching a video. Pictures form in your mind. Many healers train their eyes to see auras (energy fields). This is clairvoyance.

# CONVERSATIONS WITH PETS

### 🐾 Clairaudience (soul-to-soul)

When you start to meditate, you might be afraid of the dark. You close your eyes and everything around you is black. You see nothing, you hear nothing. But after a while you will get used to the dark and start to hear sounds. The darkness becomes a symphony, and you might hear the angels singing. The darkness becomes alive with beautiful sounds. You become aware that there is a whole world beyond the darkness. Clairaudience is when you hear these sounds.

When I was a child, I used to hear footsteps of people walking around in the house at night, even though there was no one there. In this way, you can learn to hear the thoughts of the animals.

### 🐾 Clairsentience (heart-to-heart)

Clairsentience is to feel and sense things. An animal walks up to you and you sense there is something wrong with it. You visit a friend and just know there is something they want to tell you, even though they haven't said anything yet. At times, for no apparent reason, you worry about someone and need to contact him or her to make sure all is well. You may also see someone and just know that they are unhappy and need help in some way. You are using heart-to-heart clairsentience for all the above examples.

By developing all three of these senses, you will find it easier to communicate and have telepathic conversations with your pets and other animals.

One of the questions I'm frequently asked is, "Do animals understand any language?"

My answer is a resounding "Yes." The reason for this being that they are not responding to the words you say, but the pictures that are forming in your mind. They see the intention behind the message.

Animals are not here just to be our pets. They have a much bigger purpose. They are our teachers and they teach us the true meaning of love, respect, compassion and trust. They help us to get to know our inner selves. And, although they are here to teach us, they are also here to learn from us.

Communication implies a two-way conversation and not a monologue. It's not only important to talk to the animals, it is just as important to listen to them. Part of communication is knowing and understanding their body language.

If you meet a person and they keep their arms folded, the message you get is that they are closed and not open to the conversation. If they keep their eyes down so as not to make eye contact, the impression you get is of shyness and insecurity.

Animals also have body language. You need to get to know their body language to make it easier to understand their communication.

When a dog wags its tail it is happy or pleased and having fun. If, on the other hand, its tail is rigid, it might be ready to fight. A cat wagging its tail is saying, "I'm uncomfortable, annoyed or angry". If its tail is relaxed and you can hear purring, it is happy and content. If a horse pins its ears back against its head, it is saying, "I'm angry, keep away". If its ears are pricked up and facing forward, it is alert and interested in its surroundings.

Our body language also affects animals. When animals are tense or frightened (e.g. during a thunderstorm), your body language can help to calm them down. For example, if you walk into a house and a dog comes to greet you, but you notice that it is tense and insecure, then give a big yawn and lick your lips. This has a calming effect, and lets the animal know that you are no threat.

It is very important to remember to always stay relaxed while communicating with animals. Don't expect anything, just sit, ask the question, and then wait. You will be surprised at what comes through. It is a good idea to have a pen and paper handy, as the information comes through very fast. Don't analyze the information but just accept what you get. You will make sense of it later.

Always allow the animals to give you their point of view. When I work with an animal, the people always ask me how I heal their emotional problems. Sometimes it is necessary to do a full healing session but in a lot of cases, simply allowing the animals to tell you their side of the story is all they need. Think of a person. You have stress that builds up slowly over the months and eventually you go and see somebody, be it a therapist or a friend. You tell them all your problems and get everything off your chest. How much better do you feel? The same applies to animals.

## Mishka's job

A friend of mine named Wynn asked me to talk to her Siberian Husky, Mishka, who was digging holes all over her garden and killing the birds. I went over for tea and while Wynn was pouring the tea, I quietly connected with Mishka. What I found out from Mishka was that the birds were intruders and she was doing her job by getting rid of them. I explained to her that the birds were beautiful and part of the garden. I asked her to please allow them to live in the garden with her. I also explained to Mishka that I knew she was a dog and needed to dig, but please try to dig in one spot in the corner of the garden so as not to destroy the entire garden.

# CONVERSATIONS WITH PETS

Four months later, I found out Mishka had one hole in the corner of her garden and hadn't killed any more birds.

When starting to communicate with your pets, you might go through some fear, anxiety or insecurity. It is a good idea to point out to the pets that the fear you might be feeling is a fear of your own inabilities. Are they going to answer you? Will you be able to do this? What are they going to say to you? It is not a fear of danger. Sometimes you pass on this feeling of fear and your pet misinterprets it and, thinking there must be danger, starts to panic. This could lead to them not communicating with you.

Humans often let their own perspectives get in the way. They feel animals can't be happy in small spaces such as cages or enclosures. Animals on the other hand don't judge, and are more ready to accept their situation. Some of them have chosen their particular situation, for whatever reason.

When visiting zoos or places where animals are kept in confined areas, remember that animals can and do pick up your feelings. That is why it is so important not to go in with the thought of "Shame! Poor animals!" For one, the animals themselves don't like it when they pick up your pitying thoughts, and for another, some of these animals have chosen this life for themselves. Most of them don't know any other way of life and are happy. The minute you feel sorry for them, they are immediately dissatisfied and unhappy. What you should be saying is, "Thank you so much for allowing us to learn from you, by observing you. You are so lucky to have such a beautiful home. You have a warm place to sleep and food in your tummy. Thank you for being there for us." The animal will then feel totally happy, satisfied and very content.

While animals do have a sense of humor, they always tell the truth. During one of the first workshops I gave, we asked three of my miniature horses who their favourite person is. All three answered,

# CONVERSATIONS WITH PETS

"Thomas." He is the groom who looks after them and feeds them. Shock! I thought I was their favourite! I decided not to ask my canine friends that same question. I chose to remain their favourite friend in my eyes. If you don't want to know the answer, don't ask the question.

People may misinterpret what their animals say. You may ask who their favourite person is and they may answer that it's the little boy next door. This does not mean they don't love you, but rather that they enjoy the fun time they have with the boy next door.

If you ask what your pet's favourite food is, they may give an answer you don't agree with. If they say roast chicken but you have never fed them this, you may not be aware that a friend gave them a piece at some point. They could also smell it while cooking and decide it's their favorite.

Animals love the chance to tell their side of the story. Think of spending time with a friend (human that is) who spends the time talking about himself. You don't get a chance to get a word in. How does that make you feel? It's the same with animals. Calling your animal over for a chat does not mean that you do all the talking. Allow them the opportunity to have their say. They may have interesting things to tell you. The word "communicate" indicates a two-way conversation.

Understand that being able to communicate does not mean they will automatically listen to you. They are individuals with minds of their own and need to be respected for this. Like children, they don't automatically follow instructions.

Animals need to be respected for who they are. Just like us, they also need discipline and there is nothing wrong with that. However, when you start to take away their individuality, by not allowing them to be who they are and not allowing them to express themselves, then you start looking down on them. You become the ultimate boss, you start playing God, and you don't allow them to dig, or even get excited when visitors arrive.

Sometimes you even go so far as having them de-barked because you don't like the noise. How else can they express themselves if they can't bark or, as in the case of my Huskies, sing? There are many nights that I sit on my lawn with my two Huskies and we all howl (sing) together. It's a lot of fun. Sometimes there might be a medical reason why a dog would need to be de-barked, but before making the decision you need to think very carefully of all the pro's and con's. How do you feel if you lose your voice for only a day? You can't call out, even if you are in danger. If your dog got himself stuck in the fence or injured in the garden, how would he alert you in his hour of need?

Humans have this need to be in control of everything. We feel we don't have enough control of our own lives, so what do we do? We try to control the animals that share our lives – when, in actual fact, we should be treating them with respect, as partners and best friends, and sharing some of the love and protection that they so freely give to us.

### 🐾 While I was growing up

I've always had a very strong affinity to animals. As a small child when our dogs used to come into the house, I always knew exactly what they were thinking or feeling. I thought that maybe because I knew them so well, I could read their body language. But then I realized I was doing it with the neighbor's dogs, as well as strays that came around. I didn't realize that other people weren't doing the same.

I started to play games with the dogs. I would sit quietly and think what I wanted them to do, like fetch their balls or run outside and play, and they always responded. I would think, "Wow, what clever dogs." I then started experimenting with other animals, such as horses and cats. The experiments proved just as successful, and we were having lots of fun. And our bond was deepening.

Many years later, my mother gave me a book called "Straight From The Horse's Mouth" by Amelia Kinkade. I couldn't put this book down! I thought, "Fantastic, this is what I've always done, here is somebody who can teach me how to control and direct what I've been doing all my life" – talking to animals.

I started e-mailing Amelia and asked if she was planning a trip to South Africa. She said she would love to but at this stage there were no plans. However, she suggested that, as she was running a course in the Isle of Man in a few months, I should join her there. I was so excited and I booked straight away. Dragging my mother along with me, I was off to the Isle of Man to meet one of the world's leading animal communicators.

Little did I know that, after doing a course in animal telepathic communication with Amelia Kinkade, my life would change forever. I would finally be able to give something back to the animals that had given me so much all my life.

A few months later, I started running my very own workshops in animal telepathic communication and healing. This is when I decided to donate a portion from every workshop to an animal charity of my choice every month. In this way, I would be able to help many more animals in need.

###  Riff-Raff

It was a hot sunny day and I had spent the best part of the day with my horses. I was hot, tired and ready to take a break. As I stepped in to my TV room, I sat down with a sigh. I looked up and in front of me were all my canine companions: Snoopy and Daisy, the Jack

# CONVERSATIONS WITH PETS

Russell terriers; Maggie, a beautiful Bull Terrier/Labrador cross; and Stacey and Riff-Raff, my two very mischievous Siberian Huskies. They were all fast asleep on the couches. I took one look at this lot and said (telepathically), "So guys, I see you're having a hard day at the office today." With that, Riff-Raff looked up straight into my eyes and I heard the words in my head… "Don't be ridiculous"…after which he turned over and went back to sleep.

## Mona Lisa

This story comes from Carol FitzPatrick, a friend of mine who has attended my workshops more than once. Carol takes up the story.

"I got Mona as a present from my husband, Theunis. I was allowed the pick of the litter but when I got there, ready to choose a puppy, one of them ran up to me. It was the puppy I would later call Mona Lisa, and she had obviously chosen me. We bonded very closely, she was an adorable puppy and we loved her very much, but would later discover that she came with many problems.

"At eight weeks old, we discovered she had an allergy and, after many months, we ascertained that it was an allergy to the grass. We later realized that wasn't the end of her problems. I noticed she couldn't get in the car by herself and this worried me so much I spoke to my vet about it. After various tests, the vet said there was something wrong with her hips but could not X-ray her until she was seven months old. We did the X-rays when she was nine months old and discovered she had fourth-grade hip displasia. Hans, my vet, said that normally he would suggest having her put down, but he knew Theunis had given her to me to take my mind of his illness. Theunis was suffering from cancer and had got Mona for me as a companion. The vet decided to remove Mona's hip and see how she went before making a drastic decision. The muscles took over and soon she was running around, but still had a certain amount of pain. We had to do a second hip operation, which took much longer to heal than the first one because Mona kept tearing themuscles and reinjuring herself.

"By November 2003, I was very tempted to have her put down because she was suffering. But Mona, being a very determined dog, would not give in, and she now runs around playing happily.

"She is my hero. In January 2004, Mona saved me from an attempted hi-jacking. As the hi-jackers approached me in my car, Mona leapt at them from the back seat and chased them off. For a dog without hips, it was an amazing feat. Mona saved my life and I will always be indebted to her, my special dog.

"I have been communicating with Mona for a long while now. During these sessions, she has told me she is very glad we saved her from being put down as she still has things to do in this life. She sees herself as a communicator and teacher and says she has lots of information to share with us.

"She believes she is here to look after me and does this very well. She does not see herself as a dog but as mom's companion. She just loves to be with people all the time.

"She enjoys being part of Jenny's workshops and communicating with all the people who attend. She is anxiously waiting to talk to Jenny, as there are some things she would rather tell her for fear of hurting me. I respect her for this, and Jenny said she would get in touch with her soon."

After listening to Carol's story of how she got Mona Lisa, I decided to see what Mona wanted to talk to me about. I connected to her through a photograph Carol had left with me, and this is what Mona said.

"I'm worried about my mom's health. She needs to take more time for herself and relax. She is carrying the stress of the whole family on her shoulders.

"Tell her not to worry about me so much, I am fine.

"My mom needs to take back her power and do what is right for her and not what others tell her is right.

"The problems she is having with her back relate to her taking on every one else's responsibilities.

"It's not that I don't want to hurt her feelings, but rather that I don't think she will believe this message is coming from me. She sometimes doubts her abilities. This is because she is human. It's understandable.

"I don't judge her, she shouldn't judge herself. Love is an unconditional emotion. I love her as she loves me; she needs to learn to love herself more.

"My special message to my mom (Carol):

*Always speak your truth,*
*Don't be what you're not,*
*You are enough,*
*Let your light shine through.*
*We appreciate you*

*I speak for all the animals"*

I was left feeling, what an amazing dog. I am so honored to be able to use her story in this book and share her thoughts with all the readers of my book.

# CONVERSATIONS WITH PETS

### 🐾 Riff Raff and our book

I went to visit a friend, Alba, a publisher, who later assisted me with lots of advice regarding the design, layout and production of my book.

She gave me heaps of information on how to go about writing this book. I realized if I wanted to get it out by the end of the year, I would have to put in lots and lots of work. She told me about the layout, the content, the photos and much more.

This was not going to be easy, what with all my private readings, workshops, healings, courses, articles for various magazines, and I still needed time with my own animals. How was I going to do it?

Later that day when I arrived home after my meeting with Alba, I walked into my house and all the dogs came to greet me at the kitchen door. I said to them… "How am I going to do it? Where am I going to get all the information to put in my book?"

I felt desperate, when Riff-Raff walked up to me and said, "Don't worry, I'll help you write your book."

So now I needed to find a quiet place for Riff-Raff and I to sit and start writing the book.

A few days later, Riff-Raff and I sat together in a quiet place to work on the book.

The first question I asked him, was, "Raff, what would you like people to know?"

He thought about it for a while then replied, "People are very misguided in their priorities. They think of themselves and the future instead of living in the moment. They criticize others and they lack self-love, and if you can't love yourself, how can you love others? Humans judge, they need to display more unconditional feelings. Make no mistake, they can be fun, but also too serious. People need to be more aware of their daily goings on. They need to open their hearts and relax more."

I then asked him, "How do you see other animals?"

He replied, "Intelligent, fun beings, here to teach people and lead them down certain paths in order to have certain experiences. The animals are holding this planet together, through their love, wisdom and acceptance." I asked him if he had anything more to tell us.

"People can't process too much at a time. So, take in these words, and we will chat again soon."

I thanked Riff-Raff for this information as he got up and walked off into the garden to play.

Six months had passed since Riff-Raff gave me his first bit of information for my book. He started telling me that it was now time for my second message. A few days later, pen and paper in hand, I went in search of Riff-Raff to see what else he had to say. He was lying outside in the garden in the sun. I walked up to him and sat down next to him on the grass. He sat up and looked at me, he was ready to talk.

I said, "Riff-Raff, what is it you want to talk about?"

Without delay he said, "Let's look at the word respect. What does it mean?"

He went on, "In order to successfully connect to any animal, it is important to have respect for yourself. If you have no respect for yourself, you can't possibly have respect for the animals in your lives.

"Without respect there is no love and without love there can be no connection."

"All we ask is that you allow us to be who we are, without judgment. Allow us to have our own thoughts, feelings, likes and dislikes. Allow us to be who we are and not who you want us to be – therefore showing us respect. This will give us a much more loving and fulfilling relationship."

"Be prepared to hear our voices and learn from us. We can learn a lot from each other. Think of us as equal, don't look down on us. We are your partners and friends. Look after yourselves; respect the bodies you live in. See the joy life has to offer."

"The more you communicate with and listen to the animals in your lives, the deeper your bond will become."

I thanked Riff-Raff for these amazing words of wisdom and in turn he thanked me for valuing his opinion.

As I got up to leave, Riff-Raff let out a big sigh, lay down on his side, closed his eyes and fell asleep.

### 🐾 Information from Snoopy

I said to Snoopy, "Snoo, what question do you want me to ask you?"

Snoopy answered, "How can I know what you want to ask me if you haven't asked me yet?"

# CONVERSATIONS WITH PETS

"Okay then, what are your feelings about humans communicating with animals?"

"It's about time. There is a consciousness taking place. With increased awareness comes enlightenment. Our spiritual growth depends on our awareness. There are many animals out there that are not aware of their spirituality. We need to raise their consciousness to a level where their awareness is heightened. It is the same with humans. Until someone puts them on their path, they are not aware there is a path."

"Snoopy, what is your role in raising the awareness?"

"On a soul level, I am there to give support and encourage. Unlike humans we animals learn from each other. Cats, for instance, are very relaxed and know the limits of their bodies. They teach us to focus and be flexible in our thoughts. There is much we can learn from cats. How to still our minds, for instance. Dogs generally have a great respect for cats. They teach us to love ourselves, for how can we love who we are, if we don't know what we are?"

"Try being a cat for a day. Sleep deep, on wakening, stretch every muscle from the tips of your toes to the top of your head. Feel the love move through your body, enjoy being you."

### 🐾 Daisy's input

I asked Daisy this question, "Daisy, how do you see yourself?"

This is what she said: "I am love in its purest form, look at me and learn from me. Many humans think of us animals purely as pets. What they don't realize is that we are their teachers, and we have great wisdom. If you look at us you will learn from us. We are the bearers of great joy and love. Follow us and we will show you the way."

# CONVERSATIONS WITH PETS

I asked, "What makes you so wise?"

Daisy replied, "I've been doing this for many centuries. It's my purpose in life to make others feel love. Humans are destroying the planet through their anger and ignorance. If they could feel the love in their hearts, there would be hope for all of us."

## 🐾 Riff-Raff and the "Here" command

I have a certain command that I've taught my two Huskies at school. When I call them to me, instead of saying, "Come", I always say "Here". Well, I had just taken all the dogs for a walk to the stables and I called them to come through the gate on my way back to the house. As I turned around to say "Here", so that I could give them a treat, I noticed Riff-Raff had run in ahead of us.

He was standing at the corner of the house saying, "Here. Here."

I looked at him and said, "You've got to be kidding. I'm not coming there, if you want your treat you'll have to come here." Reluctantly he came for his treat.

## 🐾 How easy it is for animals to misinterpret our messages

It was summer, Riff-Raff was moulting and looking a bit shabby. I decided to brush him to get rid of some of the loose bits of fur. I walked up to him and said, "Raff, come here and let me brush all your hair off so that you will be a lot cooler."

With that, he ran off as quickly as his legs could take him. I wondered what I'd said, why he didn't want me to make him cooler. Every time I came out with the brush, he would run off and complain as loud as he could with whining sounds.

Many days later, I was talking to him and I asked why he didn't want me to brush him. The picture he showed me was very scary.

# CONVERSATIONS WITH PETS

There was Raff with no hair at all! He was totally bald. He had mistaken my message and thought I had said I would remove all his hair and leave him bald. He is still not sure, even after I explained it to him, that I would remove only the loose bits.

It is so easy for animals to misinterpret messages from their human friends. Often the picture that we send is not what we intend to send. When I told Riff-Raff that I would remove all his hair, I was talking as though to a person. He took this literally and got the message that all his hair would go. Be very careful how you phrase a question. Take time to study the reaction you received and understand why or how your question was perceived in that way.

## Dogs and classical music

It was a Saturday and I had been taking part in an all-day workshop. While I was out, I decided to leave the radio on the classical music channel for the dogs. On my way home at the end of the day, I received a phone call inviting me out to dinner as a surprise for a friend of mine. I rushed home, checked the horses, fed the cats and dogs, got changed and left again, forgetting to change the music channel.

When I arrived home at 21:30 all my canine friends met me at the door.

They were very angry.

"Don't ever do that to us again, couldn't you have changed the channel, we have had enough, we want something different, don't go out and leave us with that music ever again."

I said, "Sorry, guys" and switched it off.

## Vista

I recently had a conversation with Vista, a cute little pot-bellied pig who lives down the road.

When I told her that she was the cutest little pig, she told me that she was not just a pig. She was a soul with a heart of gold and lots of love to give.

She said people should learn from her as she was very intelligent and had a lot to teach. She loves people even though they are a bit slow at learning their lessons.

She said, "Wouldn't it be great if we wrote a book where the animals told all the stories?"

Little did she know.

## 🐾 Bella and the workshop

One day, Vista came to join us on one of our workshops. We tried to get Bella, her companion pig, to join us but no luck. There was no way Bella was going to come to a place she didn't know. So Vista had to come on her own to chat to the people. It was a great success. Vista chatted to every one, telling them all about Bella, what she liked to eat, her favorite activity, and all about her life on Blue Saddle Ranches. She had a lot of fun, as did we. It was then time for her to go home. We all thanked her for coming and for sharing her stories with us, said good-bye, and sent her home with a packet of Marie biscuits.

A few weeks later on a Saturday morning, I was getting ready for another workshop, when the bell rang. I went to open the gate to let the first person in, when to my astonishment in trotted Bella, saying, "Vista told me about the workshop so I'm here! I'm here for the workshop." She had walked one kilometer all by herself to come to the workshop. I don't know if it was the workshop she wanted, or the packet of Marie biscuits Vista had told her about. So we chatted to her for a while, gave her the Marie biscuits and opened the gate. Off she went to tell Vista about her experience at the workshop.

Bella and Vista have become great friends with all the other animals on my property and can be regularly found chatting to the horses through the fence. Sometimes they pop in to see me for a chat and a Marie biscuit.

One year after Bella and Vista's first visit to my home, they had babies. I now have one of Bella's babies and one of Vista's babies living with my growing family and me. I have named them Merry and Pippin.

### The Alpacas

Alpacas are part of the camelid family – along with camels, lamas, vicuna and guanacos. They originated in Peru, Bolivia and Chile and were first imported into South Africa by Dr Gavin Lindhorst who is based in Cape Town. There are currently more than 500 in the country. There are two kinds of Alpacas – depending upon the fleece type – huacayas and suris, who look like Rastafarians!

They are sheared on an annual basis and their fleece is highly prized as it is light but has good insulation, making it very warm but not too heavy. It also has no "prickle factor" and comes in 22 different colours.

They are charming creatures, full of curiosity and very gentle, unless faced with a dog. They do not particularly like dogs and in some cases neutered males are being used as "guard dogs" for large flocks of sheep to keep away jackals and other predators. Although they are herd animals and all co-habit happily, they each have their own character and individuality. There are no problems introducing new animals and the whole herd looks after the babies (crias). They love life and play together, "pronking" across the fields like magical Disney creatures in the early evening.

# CONVERSATIONS WITH PETS

Early one morning, just before a workshop, a horsebox pulled up. The ramp was let down and three amazing animals climbed out. Coco, Orion and Patricia, three stunning Alpacas, had come to spend the day, and join in with the workshops.

I took them into one of the paddocks and connected up the sprinkler (I had heard that they loved water). They ran off to play in the water while they waited for us to come and communicate with them.

As we approached them, they were very cautious, but soon settled down to chat. They said that they were light workers, meaning that they help to raise the awareness and vibration of other animals as well as humans, and were here to teach us about love, purity and gentleness.

Their sole purpose, they said, was to be a connection to creation, a vessel of spirituality, and to watch over and protect mankind. They see us humans as spiritual beings trapped in our own issues. They are highly evolved souls and are here to spread their energy.

The message they have for us humans is to open our hearts, trust and be kind to our fellow humans. And never lose the element of fun in our lives.

It was great communicating with such spiritual beings.

They were amazing.

## 🐾 Archie's words of wisdom

Archie is a beautiful big ginger cat who shares my home. I've had him since birth. He was only one day old when his mother, Mehittabel, left him with me for a whole day to go out and explore in the veldt. I spent the day bottle-feeding him. Archie is the most gentle and sensitive cat, and is now 18 years old.

One day, Archie asked me if he could be a part of my book as he had something to say. Until that day, I hadn't realized the wisdom that lay in Archie.

I sat down with him and asked, "Archie, how do you see other animals?"

He looked up at me with his big blue eyes and said, "All animals are important in their way, for instance, dogs are very down to earth, and they have ways of keeping you grounded.

"Horses are symbols of power and freedom.

"Cats show you how to get in tune with your inner essence. We all have a purpose in life. The first step to opening yourself to bigger things is to open your heart. We are here to help you do just that. All things are possible to those who desire. My purpose as a cat is to awaken you to your inner self. To help you realize there is much more to you than just your physical being."

"Archie, why did you choose this life as a cat?"

"Why not? I'm sensitive and graceful, I have much to teach. Every animal comes into your life for different reasons. We are all here to teach different lessons. My lesson is to teach you to find your inner peace. Be contented with who you are, stop trying to be who you're not."

"Discover the beauty in you, and spread the knowledge."

## Stacey's message

My magnificent blue-eyed Siberian Husky, Stacey, was patiently lying in the shade waiting for her turn to give me a message for my readers.

I walked up to her and asked, "Stacey, what do you feel about your role as a dog in the family?"

Without hesitating she answered, "I love it; you should try being a dog for a day."

"Dogs live inside their hearts, everything they do is through love and motivated by love. We live in the now, not worrying about tomorrow. We love everything we do. Our meals, our games, our walks, even just lying in the sun is a loving experience for us. Every experience we have is limitless in its ecstasy. When you, as humans, feel butterflies and excitement for a new project, that is what we feel about our every experience."

I then asked her, "Stacey, what is your job and your life purpose?"

She said, "My job is most definitely to help people to experience what unconditional love is all about, and to draw their awareness to it.

"My life purpose, on the other hand, is to teach people to have more confidence and to trust themselves on a much deeper level. One of man's biggest hindrances is his lack of trust and love of himself. That is why I chose to come and live with you. So I could help to teach people to communicate with other species. People's minds are too full of self-doubt and worries about tomorrow. They need to learn to quieten their minds to be able to hear the thoughts of the animals around them. They will be far less lonely and self-absorbed. With knowledge comes power. Allow yourself to draw on the knowledge being offered to you by the animals, and increase your power."

# CONVERSATIONS WITH PETS

### Ginger's chatter

Before Ginger came to live with me five years ago, he lived with a neighbor. Every time I went to visit my friend there, Ginger would be this cute little fluffy ginger cat being dragged around by the dogs. One day, Ginger disappeared. Everyone looked for him, no one found him. They all thought he was dead.

Two years later, I heard a noise at my back door and went to investigate. There was the scruffiest, dirty, skinny ginger cat with an ear infection. I fed him and treated him, and took him to the vet to fix his ear. Every night, he would go out and fight with all the cats in the neighborhood and come back in the morning with new wounds. After a few months, I took him to be neutered in an attempt to stop his roaming and fighting.

Up until then I hadn't recognized him. My friend came for a visit and there he was all fixed up. She said, "Ginger! That is Ginger, my cat. Can I take him home?" I knew if she took him back the dogs would injure him so I said, "No, he has chosen to come and live with me."

I decided to ask Ginger to contribute to my book.

"Ginger, what have you got to teach us humans?"

He replied, "There is plenty you can learn from me. I am a very wise cat. I am thought of as an elder, or superior cat, by other cats. What I am here to teach you is how to expand your horizons. Because you are human, it doesn't mean that you cannot feel what it's like to be another being.

"I have lots of work to do. You humans are very closed to your abilities. Because of this, you limit your own potential.

# CONVERSATIONS WITH PETS

"I am here to demonstrate and lead you down a path of imagination, for it is ultimately your imagination that will assist you in communicating with your animal friends. If you can't imagine what we are saying, how will you be able to hear our thoughts? I will teach you to be confident and strong in your beliefs.

"Animals have been communicating with humans for many centuries. But because of the mundane world in which humans live, they have not been open to or able to hear the messages. People need to laugh more and relax. They are too serious.

"Relax, have fun! We will still be there tomorrow."

### 🐾 Frodo's story

It was a Saturday night, the beginning of winter and the weather had gone cold. I decided to go to bed early.

At 22:30 the doorbell rang and I wondered who it could be at this hour. I went down stairs to see. When I opened the door, there was Ernest, one of my neighbor's gardeners. He had been on his way over to visit Thomas, my gardener, when he heard a kitten meowing in the field. He stood in the road for about 45 minutes coaxing the kitten out of the field. He picked the kitten up and brought it to me. Now as I looked out, here he was at my door with a six-month old beautiful grey kitten.

Ernest and I brought the kitten in and took him to one of my spare rooms where the dogs wouldn't worry him. While I fetched some food for the kitty, Ernest organized a sandbox. The kitty was very hungry, he ate all the food, and then jumped straight onto the bed and went to sleep as if he had always meant to be here and had finally come home.

# CONVERSATIONS WITH PETS

I placed notices up everywhere to see if his people were looking for him. No one came. Then one day as I walked into his room I heard him say, "I want to live here with you."

I immediately went and took all the notices down, two weeks had passed already. I decided to name him Frodo. Frodo was here to stay, having indeed come home. From the moment Frodo entered my home, he was totally comfortable and looked as if he belonged.

As I write Frodo's story, he has just climbed up my back and is sitting on my head, from time to time looking down into my eyes from above and gently tapping my forehead. He, too, has a message for this book.

## 🐾 Savannah

I was asked by Judy Geyer to please chat to her beautiful, gentle pitbull Savannah.

One day, Savannah came running up to Judy as she got home from work. Her one hind leg was bleeding, the bone was sticking out and the artery had been severed, there was blood everywhere. Her leg was obviously broken. Judy was horrified and rushed Savannah to the vet. She was operated on and Judy took her home. Judy took time off work to nurse her baby. After two weeks of constant nursing, her leg was rotting and Savannah was in a lot of pain. The leg had to be amputated. Judy loved her dog and was totally distraught. Her baby was about to lose a leg, how would Savannah cope as she was so active and energetic? What would she do? Was she going to be alright? Judy was heartbroken about the possible effects this might have on her beautiful, energetic dog's life.

Many months later I sat down to ask Savannah what she felt about her ordeal. This is what this brave dog said.

"I won't lie. During the procedure I was afraid and anxious and felt some pain. My family looked after me and I was grateful for that. Humans always put too much emphasis on physical looks and not enough on the soul. Animals don't see the physical body, but rather the soul. We love with all our hearts, unconditionally. To us it doesn't matter if you have one leg or six legs - we see your soul.

"Just because the physical body is damaged, it doesn't mean that the soul is also damaged. The soul is still whole. I am alive and happy. My mom has accepted me as I am, and this makes me very happy. She feels guilty but she shouldn't. It was no one's fault.

"Some humans feel it is better to put an animal down, rather than amputate a limb. Doesn't everyone have a right to live and be happy? It does take some adjusting, but with the loving support of our human family, this is easier.

"All animals deal with amputations in different ways. The way humans can help them is to give them lots of support and love, and help them through the pain with some pain medication. Accept them unconditionally.

"My message to all humans is:

Open your hearts to your animal friends.
Accept them unconditionally no matter how many limbs they have.
Treat them with kindness and respect.
Allow them to bring joy into your lives.
The animals in your lives understand more than you could imagine."

# MEDITATION
## 🐾 Removing the cloak of negativity 🐾

Sit in a quiet safe place
where you can relax and won't be disturbed.
Uncross your arms and legs.

Focus on your breathing and the sound of your heart beating. Feel every muscle in your body start to relax.

You find yourself standing in a field,
you are dressed in a heavy dark black-hooded cloak. Feel. Feel the coarse texture of this cloak. This cloak is made up of all your negativities, all your anxieties and insecurities. Feel the weight of this cloak on your shoulders, pressing down. Now, slowly, you feel this cloak start to lift, it moves up, away from your body, off your shoulders and head, taking with it all your negativity, insecurity and anxiety.

You are left with a feeling of lightness, joy, happiness and excitement; all your negativities and tensions are gone. You notice, in the distance, a fountain rising from the earth, it is a brilliant white fountain of light.

As the white light rises from this fountain,
you notice it comes cascading down from above as hundreds and thousands of tiny little silver stars.

You become aware that you are now cloaked in a different cloak, a beautiful translucent cloak, woven from the stars. You have a feeling of excitement, confidence, peace and happiness. All around you are rainbows of the most beautiful colours. Above you is a ball of white light.

As you watch, it slowly starts to form a
circle of protection around you, leaving you encased in a beautiful bubble of white light.

You start becoming aware of your breathing, and the sound of your heart beating. Listen to the sounds around you. Feel the room around you, start moving your fingers and toes.

Then, slowly, open your eyes.

# EXERCISE 1
## 🐾 Quieting your mind 🐾

Sit in a quiet place either in your house or in the garden. Take some deep breaths to reach a relaxed state.

Start to allow your thoughts to drift out of your mind.

Don't hang onto your thoughts, rather acknowledge them, then let them go.

If it helps, see them on a cloud drifting away.

The more you worry about your thoughts, the less chance you have of quieting your mind.

Once you are relaxed and your mind is quiet, then start to hear all the sounds of nature around you.

This is not as easy as it sounds but will require some practice.

It might take you a week or a month, but keep practising on a daily basis until you feel ready to move to the next exercise which will appear in the next chapter.

## CHAPTER Three

## THE EXTRAORDINARY IS ORDINARY - X-ray vision and body scan

*Look at the animals in your life,
and ask, why have you
chosen to come and live with me,
what is it I can learn from you?
And what is it you can learn from me?*

*Respect them and experience being unconditional.*

*For they don't judge you,
give them the same courtesy.*

**Snoopy**

*Trust in the process; get out of your own way,
Open your hearts and live life to the fullest.
You are worth it; you just don't know it yet.
I pass my love on to you, receive it and enjoy it.*

**Stacey**

# THE EXTRAORDINARY IS ORDINARY

Centuries ago, if you were caught doing spiritual healing or psychic work, you were burnt at the stake. All psychic work was thought to be evil and dark. Thankfully today many more people are opening up to the possibility that there really is something to it, and that animals can communicate with people. People can heal their animal friends, and animals really do have souls and go to heaven when they die.

The crux of animal communication should be for us to improve the lives of our animal friends, and not to prolong them without any quality of life.

"Are you in pain?" is one of the most important questions you could get an answer for from your animal friend. Imagine suffering from headaches, arthritis, backache, depression, anxiety attacks or any number of ailments that are not life threatening. What would happen if you couldn't explain to the doctor what you were feeling? He wouldn't be able to effectively treat you.

Very often animals have this same problem since they cannot explain how they feel. This is where the communication comes into place. You shouldn't have to wait for your animals to be on death's door before you notice anything is wrong. You should ask every day how they are feeling, and hear their answer. This way you will be able to effectively treat them.

## The Gestalt method

Gestalt is a technique where you put yourself inside an animal and feel what it is like to be this animal. Feel its fur growing on your face, what can it smell, how does it feel emotionally? Look at the world from this animal's point of view. What do we humans look like to this animal?

Gestalt is nothing more than a concentrated form of creative

# THE EXTRAORDINARY IS ORDINARY

visualization. It's like a game we all played when we were children and we pretended to be something else. If you pretend to be an animal, you can identify with a myriad of physical complaints. It is so simple that most of us have forgotten how to do it.

Think of it as a game.

Your biggest challenge will be to trust what you get. Before attempting to get inside your animal's body and identifying the feelings of your animal friend, first get inside your own body and identify your own feelings.

The key word here is trust. You need to trust everything you feel, see or sense. You need to have an absolute awareness of what is going on in your own body, so that you can differentiate between your body and that of your animal friend.

With Medical Gestalt, you place your consciousness inside the animal so as to travel through its body to find health problems and to determine any damage or disease to the internal organs, the cells, muscles, bones, and to look for any impurities in the blood stream.

The Gestalt method is the only effective way to search from the inside to find painful problems inside the animal's body. It is important to realize, however, that although this can be very beneficial and of great assistance, it is not a replacement for traditional veterinary care.

Then there is the "remote viewing" method of tracking an animal. This is where you take your consciousness to where the animal is, no matter how close or far away it is, in order to see what it is doing. At no time do you physically leave your body.

# THE EXTRAORDINARY IS ORDINARY

As with any form of communication, it is very important to ask permission before connecting to animals, especially in the case of Gestalt. It is an invasion of their body, and they need to be happy to allow you in.

Before talking about how to do Gestalt and remote viewing, here are some stories on trusting the messages you get when communicating with animals.

### 🐾 Horse with an abscess

The names in this next story have been changed to protect the privacy of the people involved.

A friend of mine, Mary, had lost her beloved horse a year ago. It was devastating for her and had taken her a whole year to decide to get a new horse. She eventually made the decision and bought Gambit, a beautiful dark bay gelding.

After having Gambit for three months, he became lame and was taken to the vet. On further examination it was determined Gambit had a fracture to the pedal bone in his hoof. They treated it with antibiotics to prevent infection, gave him painkillers and also applied a poultice in case of an abscess.

When the vet later did another check, she said there was no evidence of an abscess and couldn't understand why he was still so sore on the other side of his hoof. They decided to do some X-rays the following week.

In the mean time, Mary had asked me to see if I could pick up anything by doing a Medical Gestalt and communicating with him, which I did. I found a massive deep-seated abscess in his hoof. The lamina had separated from the wall of the hoof and the pedal bone had shifted.

# THE EXTRAORDINARY IS ORDINARY

I said to Mary that although we knew the vet hadn't found an abscess, I was picking up a big deep-seated abscess, as well as a condition similar to laminitis where the lamina and pedal bone are affected. I asked her to see if the vet could do another check.

When they got Gambit to the surgery, the first thing they did was remove his shoes then went ahead with the X-rays. The vet was horrified to find a massive deep-seated abscess. The lamina had ruptured and the pedal bone had shifted.

They now knew what was wrong and could treat it effectively. Gambit was instantly happy to know that his problem had been found and he was now on his way to a healthy recovery.

## 🐾 My spiritual wolf

For two years I have had a feeling that there was a wolf around me wherever I went. When I drove home at night along the quiet road near where I live, I would get a very strong feeling of a wolf in the trees lining the road. I even drove more carefully, expecting to one day see my wolf standing on the side of the road.

One day while meditating, I was finding it difficult to concentrate. My mind was too busy and so, after about 15 minutes, I decided to stop and try again later. As I opened my eyes, I saw this magnificent wolf run from the kitchen through the lounge past the dogs and out into the garden and gone. I sat there marveling at the experience.

A couple of months later, I was running an animal Reiki course at my home. At these courses we always draw an animal card in the morning to see what messages we would get for the day. The card I drew was? That's right, a wolf.

At the end of the day, I had a photograph taken of us all standing

# THE EXTRAORDINARY IS ORDINARY

together with the horses, as a group. I had the photographs printed to give to everyone with their certificates. Lo and behold, when I looked at the photo, right in place of my face was the head of a wolf!

A wolf is a very good and strong totem to have. One of the things it indicates is deep psychic abilities and great intuition.

I feel very honored to have this wolf helping me with my work.

 Isis

I have never believed in "coincidence" and believe that things always happen for a reason.

For the last few years I have seen many litters of puppies in different places. I always LOVE to see puppies and spend time with them to just experience their joy and innocence. Many times people have said to me "Wouldn't you like a puppy? I have a puppy that would fit into your family beautifully!" My answer is always the same. "No sorry. I can't take on a puppy at this stage". Well things were about to change!

Riff-Raff had developed a small mole on his nose. I had taken him to see the vet and she had suggested having it removed. I made arrangements to bring him in the following week for this minor surgery. Whenever any of my animals have any form of surgery or procedure, I always stay with them until they are finished and can come home. This day was no different.

Riff-Raff had been taken through for his procedure to begin and I was quietly sitting in the waiting room chatting to the receptionist when a lady brought in a couple of husky/malamute puppies. These little balls of fur were coming in for their vaccinations.

# THE EXTRAORDINARY IS ORDINARY

Immediately I felt a huge pull in my heart and asked if she had any females. She said there was one female that was sitting in the car waiting to be brought in. I went straight out to the car to take a look.

I took one look at this little puppy sitting on the back seat. Our eyes met and it was love at first sight. We both just knew each other. There was such a deep bond between us and so I decided right there and then that this puppy HAD to come and live with me and join my family. I made arrangements to go and fetch her the following week. I wanted her to be a little older and I also had to tell my other dogs about her before she arrived. They needed to think it was THEIR decision to bring her home. This would help us divert any potential jealousy issues. It was the most difficult week of my life. Waiting to go and bring "Isis" home.

The day finally arrived and off I went to pick up this special little lady. My mother came with me so that she could drive while I held Isis on my lap. We got to the ladies house and saw all the littermates were playing and running around quite confidently, however Isis was VERY timid and was hiding. The minute I had her on my lap in the car and we were on our way home, she sat up and became perky and totally happy. She just knew that she was going home.

When we got home I let her out of the car as I wanted to give her some time to settle before introducing her to Riff-Raff and Stacy – it only took a matter of minutes. Riff-Raff and Stacy were creating so much noise at the fence that I let them through. Stacy being the alpha dog went straight up to Isis and sniffed her, welcoming her to the family. When Riff-Raff came up for his introduction, Isis took one look at him and fell head over heels in love with him. From that moment on she followed him everywhere he went. She even slept with him at night. Stacy loved her but found her a little

# THE EXTRAORDINARY IS ORDINARY

sharp at times so she kept away from her, but Riff-Raff took her under his very big paw and looked after her as if she was his very own puppy.

Isis had only been with me for three weeks when I had to go to Cape Town to run a workshop. My mother and sister would be staying in my house to keep an eye on the animals, especially the new puppy. Before I left I asked Riff-Raff to please look after Isis while I was away and to take her out at night so that she could do her business in the garden and not in the house.

When I got home after my trip, my mother told me that two or three times during the night Riff-Raff would go upstairs, fetch Isis and take her out to pee. I was so proud of him for doing his job so well!

This just proves the importance and value of telepathically communicating with your animals.

Bringing Isis into our family has re confirmed to me how animals take on the behaviours of whomever they are the closest too. Isis has taken on literally every movement and characteristic that Riff-Raff has. She eats like him. Plays like him, walks like him and even sings like him.

The first day I had her I realized that there was something very different and quite special about her. This came to my attention when Rikki, the stray cat, walked in and joined our family. He was quite a dominant cat, and Ginger, the other cat was nervous of him. Rikki had grown up on the street and was a tough little guy. One day I was sitting watching television when I heard a noise. The two cats were hissing at each other. Isis jumped up and ran to the kitchen where the cats were, I followed on her heels. I did not want Isis to injure any of the cats. When I got to the kitchen, I

# THE EXTRAORDINARY IS ORDINARY

found Isis standing right in the middle of Ginger and Rikki. Every time Rikki tried to approach Ginger, Isis nudged him with her nose to keep him away. She was protecting her second best friend, Ginger.

She stood between the cats for as long as it took for them to settle and move away from each other. Over the next few years I observed Isis's behaviour around the cats and every time they looked at each other, Isis would slowly get up from wherever she was and stand between them until they relaxed.

Whenever I was busy and there was a little tension between the cats all I would do is whisper "Isis" and she would go and calm down the two cats.

## The bees

It had been a lovely hot day and I had had family over for a visit. Around mid-afternoon it was time for them to leave. I walked to the gate to see them off. The gate was quite a long way from the house, and there was a long driveway with no trees or any form of shelter between the house and the gate.

They left and I turned around to walk back to the house when I heard a deep rumbling, humming sound, getting louder and louder.

I looked up and before I knew what was happening, bees surrounded me. All around me it was pitch black with bees. There was nothing I could do. I was too far from the house to seek cover, and I was on my own.

I immediately closed my eyes and mouth, and instinctively put myself in a bubble of white light for protection.

# THE EXTRAORDINARY IS ORDINARY

I acknowledged the bees, and with that, they opened up like the Red Sea, with not one of them touching me, and I could feel the air as they brushed past me. Just as quickly as they had come, they were gone and I was left standing there reflecting on what had just happened.

## 🐾 Keli and Lynette

Two of my very good friends, Keli and Lynette, decided to go away from the hustle and bustle of the city. They went away for a few days to Keli's aunt's farm on the river, where they meditated, relaxed and spent time connecting with a bull frog they named Claude.

On their second day, they decided to go for a long walk. They had walked up the hill and down the valley when Keli heard a strange noise. She looked around but couldn't see anything. After a while they noticed the sound was coming from a bird who was flapping his wings together and making a clicking sound. They watched this with great interest and then decided to carry on their way.

The little bird kept following them for miles and, every time they turned around, there he was, flapping away.

"What is he trying to tell us?"

In the distance they saw some old ruins and decided to go and explore. By now the little bird was frantic.

"What are you trying to tell us, little bird?"

Keli got a feeling from the bird, "Don't go there!" Embarrassed, she asked Lynette, "Did you get anything?"

# THE EXTRAORDINARY IS ORDINARY

Lynette said ,"No."

She also was embarrassed to say she heard "Don't go there!" Was it her imagination? They kept walking.

When they got to the ruins they saw big slabs of concrete and wondered what they were. By now Keli and Lynette were joking about everything. But when they stepped onto one of the slabs of concrete, they went cold. They noticed there were no birds, frogs or any animals around, not even insects. There was a deathly silence. All the hair stood up on the backs of their necks and down their arms. They turned around and quickly started to leave. As they rounded the corner they started to run, they ran all the way back without saying a word.

When they got back, Keli mentioned to Lynette that she had got a feeling from the little bird that said don't go there. Lynette told her she had got the same message but wasn't sure if it was her imagination or not.

Together they lit some herbs and did a process to clear their energy. Then they went to see an old man who lived next door. He told them that many years ago this place had been a detention centre where people had been tortured and murdered.

Keli and Lynette decided to go back with a sangoma who could clear the energy and help all the souls to find peace.

The big discovery they also made was to listen to the voices of the animals and trust in what you hear. The little bird had given them a warning and they hadn't listened. Sorry little bird, they will listen next time.

You don't have to be a top class animal communicator, just listen to the voices inside your head. It is nature's way and the way of

# THE EXTRAORDINARY IS ORDINARY

the animals to communicate with you. All you need do is to start acknowledging your inner voice.

### 🐾 A quick body scan before doing Medical Gestalt

In order to do an effective Medical Gestalt, you first need to do a quick body scan on yourself to determine all your aches and pains. In this way, you will not confuse them with the feelings you pick up from the animal you are working on.

While sitting in a comfortable chair focus on your head:

Do you have a headache?

Look out of the right eye: is the vision clear or murky? Now look out of the left eye: how is your vision in this eye?

What does your nose feel like? Is it blocked or runny?

Move to your mouth, do you have a bad taste in your mouth? Do your teeth hurt? Is your jaw tense? Does your throat hurt?

Move down to your chest, what do you feel here?

Slowly move down to your abdomen. How does your tummy feel? Your hips and pelvis? Are there any aches or pains?

Travel down your thighs, knees and calves. Are there any pains or stiffness?

Move to your ankles and feet.

Now go back to your neck and travel all the way down your spine. Do you feel any abnormalities or discomfort?

# THE EXTRAORDINARY IS ORDINARY

Be aware of your lower back and buttocks.

Take your awareness to your shoulders. Move down from your shoulders into your arms all the way to your fingertips.

What are you feeling here?
Now that your body scan is complete, you can go ahead with the Medical Gestalt on your animal.

## 🐾 Medical Gestalt

Before you start, always ask the animal for permission to work with him or her. You will get a feeling of either yes, in which case you can proceed, or no, in which case you need to respect this and try again later.

Bear in mind that at no time do you actually leave your body. You are fully in control and aware of your body at all times.

This process is totally safe. All you are doing is moving your consciousness around the animal's body to determine any physical problems. As mentioned before, this is not a replacement for traditional veterinary treatment but can be of great assistance.

You can either sit opposite or next to the animal you are working with.

Or, in some cases, you can use a photograph and place it on your lap.

Look at the animal or the photograph and connect with them by visualizing a green light going from your heart center to their heart centre. Place the animal in a bubble of white light for protection.

# THE EXTRAORDINARY IS ORDINARY

Once you have made the connection, you can proceed.

Close your eyes, focus on your breathing and relax.

Drop down into your heart and picture yourself as a tiny white spark of light sitting in the center of your heart. See your heart beating; hear the sound the beating makes. Feel the love radiating from your heart.

Slowly, you start to travel up from the center of your heart to the crown of your head, where you exit.

You land on the crown of your animal's head and enter, traveling firstly through the forehead and all around the head: do you feel any tension, any headache, or do you feel relaxed and fine?

Now move up to the right ear: is the ear red and inflamed, or is it a healthy colour, does this ear look clean or dirty, what sounds do you hear, are they clear and loud, or slightly muffled? Travel down the ear canal. What does the eardrum look like?

When you have finished with the right ear, move over to the left ear and investigate this ear. What does the left ear look like? Is it healthy or dry and itchy? Listen to the sounds out of this ear, are the sounds clear or are you having difficulty hearing out of this ear? What does the eardrum look like? Do you have earache?

Now move into the right eye: is this eye red, dry and scratchy or moist and healthy? Open the right eye: what is the vision like, is it clear or cloudy? Close the right eye again. What do the blood vessels look like?

# THE EXTRAORDINARY IS ORDINARY

Move to the left eye: how does this eye feel, does it feel red, dry and scratchy or moist and healthy? Open the left eye and see what the vision is like out of this eye. Once you have done this, close the left eye again. What do the blood vessels look like? Are they healthy?

Travel to the nose. Move slowly down the inside of the nose and look at the colour, is it a healthy pink or a pale gray unhealthy colour? Is the nose dry and itchy, or damp and runny? Does the tip of the nose feel warm or cold?

Now breath in through your nose and notice, what can you smell? Do you have difficulty breathing through your nose, is it blocked or is it runny? Are the smells clear?

You are now going to move into the mouth.

Start with the upper teeth: are they clean and white, or dirty and yellow? Do they hurt or are they healthy and strong? Travel along the upper teeth looking for any sensitive areas, cavities or abscesses. Once this is done, go to the lower teeth. Do you see any cavities or abscesses? Is there any pain or any sensitive area?

When you have finished your journey along the teeth, have a look at the tongue: is it healthy and pink, or white and dry and full of cracks? Picture yourself eating a biscuit, what does it taste like? Is it pleasant or do you have a bad taste in your mouth? Look all around the inside of the mouth, what do you see? Are there any ulcers? Does the inside of the mouth look healthy?

# THE EXTRAORDINARY IS ORDINARY

Now move to the throat and swallow: is it easy to swallow or is your throat sore and scratchy? Is there tightness in your throat or does it feel fine?

Move into the chest and look at the heart: is it beating strongly and healthily or is it slightly weak? Are there any blemishes, do you see any damage to the heart muscle? Is the blood flow to the heart strong and healthy?

Go to the lungs. What colour are they? Do they look healthy or do you see any abnormalities here? As you breathe in and out, is it easy, do your lungs fill up with oxygen, or do they struggle?

Have a look at the other organs. Liver: what do you see? Kidneys: are they strong and healthy? Bladder: is it working properly? Are there any infections here?

Spend a few moments looking at the spleen, pancreas, and intestines: what do they all look like, do you notice any abnormalities, blockages or discomfort in these areas?

Now move on to the stomach: is the stomach bloated, are there any abnormalities here? What does the digestion look like? Are there any blockages?

Once you have finished in these areas move to the neck. Start to travel from the neck all the way down the spine noticing any stiffness or feelings of pain. Do all the vertebrae look healthy? Travel all the way to the tip of the tail.

Then move into the right hip: is there any stiffness or pain here or is it flexible? Take note of the tendons: are they healthy?

# THE EXTRAORDINARY IS ORDINARY

Move all the way down the back right leg and into the paw. What do the toes look like and the toenails? Do they need clipping or are they short?

Move over to the left hip and do the same. Feel and look for stiffness or pain, or are they flexible? Travel down the left hind leg and into the paw. How does this leg look? Is it fine? What do the toenails look like?

When you have finished here move to the front right shoulder. What does this shoulder look and feel like? Carry on down the front right leg into the knee and paw. Look at the toes and toenails.

Go now to the front left shoulder and feel for any pain or stiffness as you move all the way down the leg and into the paw.

Once you have done all this, jump into the blood stream and travel right through the blood: does it look healthy or slightly anemic? Is the blood flow strong around the body? Are there any impurities in the blood or does it look healthy?

When this whole process is complete, start making your way along the body and back up to the crown of your animal's head, where you exit. You re-enter your crown and settle back into your heart.

Start becoming aware of your heart beating and the sound of your breathing. Become aware of the room around you and the sounds of the birds singing in the trees outside.

Take a few deep breaths and when you are ready open your eyes.

Thank the animal you are working with and write down anything you picked up.

# THE EXTRAORDINARY IS ORDINARY

### 🐾 Gabriel joins the family

It was a very sad day for all of us when Riff-Raff crossed the rainbow bridge. His absence left a huge hole in our family. Stacey was a year younger than he was, so I felt worried about what would happen to Isis once Stacey crossed. I really did not want Isis to be alone. Because of this, I kept talking to Riff-Raff and I asked him to please send the right dog along when the time was right. I wanted someone that Isis could bond with and develop a relationship with as she was extremely close to both Riff-Raff and Stacey.

Well, after only one month, I suddenly got such an urge to go and visit the husky rescue centre. A friend of mine runs the centre and I had been wanting to visit her for some time but just did not get the chance.

I have, for a VERY long time, learned to trust my ''gut'' feeling and never ignore it. At this time I had not planned to actually bring another dog into the family – little did I know what Raff had planned?

Alan and I were sitting there with Joanne when I suddenly asked ''Do you ever get puppies in?'' She replied, ''We don't keep puppies at the centre but we do keep them in a foster home, and at the moment we have a litter of five puppies. They are border-collie husky cross''

I asked if she had a male as I felt a male would better fit into my family than a female would at this stage. She told me that they had four females and one male but he was already booked to go to

# THE EXTRAORDINARY IS ORDINARY

someone else. Not deterred, I immediately phoned the foster mom and made plans to go and see the puppies.

As we walked in to meet these stunning little creatures, there was Gabriel. He came straight up to us and he was just SO sweet, full of fun and loving. Before we had even finished visiting (unbeknown to us) some phone calls had been made to secure for Gabriel to come and join our family.

It was such an exciting time, however we still had to wait a week for him to be eight weeks old and have his first vaccination before we could fetch him. It is never advisable to take a puppy away from his mother before eight weeks old.

When we got him home we slowly introduced Stacey and Isis to him. Stacey being the alpha was a little indifferent at that time but it was love at first sight for Isis. Every time we saw Isis she had this black attachment to her neck or her tail. She was so patient with him.

I am so looking forward to seeing how Gabriel (named after Arch Angle Gabriel) develops as a spiritual and telepathic being.

Thank you Riff-Raff for bringing us this amazing little dog to keep us focused and fill our hearts with love every single day.

Gabriel was with us for seven months when Stacey joined Riff-Raff on the other side. He had gone out of his way to keep us cheerful and feeling light and happy. He is an awesome boy.

Through all my years of connecting telepathically with animals, I have realised that all it takes for our animals to grow spiritually, is for us to just acknowledge them as spiritual beings and see them

# THE EXTRAORDINARY IS ORDINARY

as the intelligent sentient beings that they are. The rest we leave up to them.

Welcome to our family Gabriel and thank you for being the joy we feel in our hearts.

# MEDITATION
## 🐾 Meeting your animal guide 🐾

This is a meditation for meeting your animal guide, who is going to help you communicate with animals.

Always remember that your imagination is an important part in developing your intuition so that you can communicate with animals and hear what they have to say to you.

Make sure to find a comfortable safe place to sit, where you won't be disturbed.

Uncross your arms and legs. Close your eyes.
Take a few deep breaths and feel yourself relax.
Let go of all your tension and clear your mind.

You find yourself walking along a beach looking out over the ocean. Listen to the sounds of the waves breaking on the shore, the seagulls up above. Look out over the vastness of the ocean. Feel the sun beating down from above, the breeze brushing against your cheeks as you walk.
Smell the sea salt as you get closer to the water.

The water is crisp around your ankles as you make your way into the ocean. With every step you feel the cleansing and calming effect of the water as you go deeper and deeper into the ocean. You now find yourself at the bottom of the ocean. You are safe.

As you look around, you see all different shapes, sizes and colours of beautiful sea creatures. The fish are swimming all around you; you see corral, seashells, and beautiful sea urchins.

*continued...*

The sand is white and soft beneath your feet.

In the distance, you see an old rusty shipwreck.
You decide to explore the shipwreck.

You climb into the shipwreck and right in front of you, there is a spiral staircase leading deep into the center of the earth. As you start your descent down the staircase,
into the center of the earth,
you count silently with me from 20 backwards.

20, 19, 18, 17, 16, 15, 14, 13, 12, 11, 10, 9, 8, 7, 6, 5, 4, 3, 2, 1,

You are now in a place where there is no time, a world between worlds, a space between space. You are completely safe.

You find yourself in the most beautiful forest. In front of you is a path. You start walking along this path. As you walk along, feeling the ground beneath your feet, hear the sounds that the leaves make as they crunch beneath your feet. Smell the dampness of the earth and the bark of the trees that are around you as you walk deeper into the forest.

The birds are singing in the trees, the sun is filtering through the branches above. The breeze is gently brushing against your cheeks as you walk. You are feeling safe, relaxed and happy.

Now ask your animal guide, who is going to help you communicate with animals, to come out of the forest and meet you here today. Your animal guide might scurry out of

*continued...*

the bushes next to you, come walking down the path in front of you, or rise up out of the ocean.

This guide could be anything, an animal you have shared your life with and who is now on the other side, a wild animal from the forest, a giant insect or even a cartoon animal dressed in human clothes. Accept this animal immediately for whoever or whatever he or she is and ask your animal guide this question:
"What can I do to help you to help me
communicate with animals?"
Listen to the answer.

Now, as you walk side by side with your animal guide along this path, take a few moments to ask him or her about your psychic abilities, the path you have chosen, and the answers to any questions you would like to ask.

Listen to what he or she tells you and remember it, for it will help you later.

It is now time to take leave of your animal guide, but before you go, ask him or her to give you a gift. This gift could be anything, a crystal ball, a ruby, a pebble, a piece of bark from a tree, a teddy bear, a beautiful gift of love to keep in your heart. Take note of this gift, it is significant.

Your guide will now give you a silver thread. Tie one end of this thread around your left wrist. The other end your animal guide keeps, and every time you have difficulty communicating with an animal, all you need to do is pull on this silver thread and your guide will be there to help you.

*continued...*

Now thank your guide for coming to meet you here today, for the gift and the messages he or she has given you, and promise your guide that you will visit him or her often. And know that every time you want your guide, all you need to do is call and he or she will be there to meet you.

Slowly, start to count silently from 1 to 20.

1, 2, 3, 4, 5,
start becoming aware of your breathing,
6, 7, 8, 9, 10,
feel your heart beating,
11, 12, 13, 14, 15,
become aware of the room around you,
and the sounds you hear in the distance,
16, 17, 18,
feel your psychical body,
19, 20,
start to move your fingers and toes.

You are back in our waking world.

When you are ready, open your eyes.

Sit quietly for a few moments.

# EXERCISE 2
## 🐾 Opening your heart 🐾

Sit with your animal friend in a quiet place, where you won't be disturbed.

Open your heart to your animal friend and feel the love pour out.

Think of the person you love the most in the whole world and magnify it ten times.

That is the amount of love you are sending to your animal friend.

Now feel the love coming back at you from your animal friend.

One way to open your heart is to visualize a green light coming from the centre of your heart to the centre of your animal friend's heart, helping you to connect to your animal friend.

Practise this every day till you feel ready to move on.

*Feeding time at the Elephant Park*

# Chapter Four

## THE INTUITIVE EXPERIENCE -
### Tracking through Gestalt

*Reach for the stars,*
*if there is something you want,*
*go out and get it, make it happen,*
*for you are in control of your life.*
*Don't let the stresses of being human slow you down.*
*Reach for the stars*
*I will help you on your journey.*

***Archie***

# THE INTUITIVE EXPERIENCE

For many years the Native American Indians have been honoring their animal guides and their ancestors. They would sit around a fire, and call upon their animal guides and their ancestors. Then they would get up and move around the fire in exactly the same way as their animal guide. If their animal guide was an elephant, they would picture themselves with a long trunk and sway it in rhythm, moving just as an elephant would move. During the course of this dance, they would go into a trance and become the elephant. This would be their way of honoring their animal guide.

This is one of the techniques we use. We first picture a dog in front of us, then slowly see ourselves as the dog, becoming the dog, feeling what it feels like to have fur on our faces, what the world looks like from a dog's perspective, how they see us humans. What does it feel like to have a tail? What does it feel like to have four paws?

Then we would try becoming a cat. We would feel what it is like to be supple and able to jump effortlessly up onto a table. What do dogs look like from a cat's perspective? Later, we move on to a lion or elephant, even a bird – we feel what it feels like to fly and soar above the trees. We experience all the emotions that go with whatever animal we have become.

All these techniques are very useful when tracking a lost animal. You can see through their eyes what they are looking at.

- Is the grass long?
- Are they with people?
- Are there any other pets with them?
- Are they stressed or happy?
- Are they injured in any way?

You can determine if the animal is still alive or has crossed to the other side. When an animal is still alive you will feel a denseness

# THE INTUITIVE EXPERIENCE

in the energy, a type of heaviness and some anxiety, stress or even pain. Once an animal has crossed over, there will be a lightness, an absolute joy, no stress or tension of any kind. The animal will appear in your mind slightly blurred and not clear. They are appearing from a higher and faster vibration, therefore giving a blurred appearance.

When tracking a lost animal, we can use all the above-mentioned methods as well as "remote viewing". This is where we move our consciousness above the lost animal to see where he is.

A good way to do this is to picture yourself as a bird flying above. This gives you a better perspective of where the animal is, as animals can only see so much from where they are.

All these techniques are totally safe. At no time do you actually leave your body. You are fully in control of yourself at all times.

Remember, the world from an animal's perspective is very different from our own. Try lying on the ground and see what the world looks like from there.

### 🐾 Tracking a lost kitty

Early one Saturday morning as I was on my way to give a talk, my cell phone rang. It was a very distraught lady. She explained that her cat had gone missing the evening before, and asked if I could please help her to find him.

I told her that I was on my way to give a talk but when I got back I would see what I could do. She needed to email me a photograph of her cat in the meantime.

Much later that afternoon when I got home, I checked my emails. There was just a description of her lost kitty, not a photograph.

# THE INTUITIVE EXPERIENCE

I was exhausted after the day's work, and wondered how I was going to track this kitty when I was so tired. I couldn't wait till tomorrow, this kitty was lost and had to get home before something happened to him. I decided to take the description of this kitty with me up to the bath where I could relax and focus on tracking him.

While I was relaxing in the bath, I connected to him and saw him under a wood-framed bed. There was a big ginger cat there, who was intimidating him and he couldn't come out. He said he was close to home and his person should go door to door looking for him.

When I got out of the bath, I phoned the lady and told her what I had picked up and that she needed to go door to door and she would find her kitty.

On Sunday morning as I was preparing for a workshop, the phone rang. It was the lady with the lost cat. She was overjoyed. She had gone door to door on Saturday night and had found her kitty, two doors away from her complex. He was hiding under a wood-framed bed and there was a big ginger cat there.

When tracking a lost animal, the first questions you always ask are:

- Are you injured?
- Have you crossed over?
- Do you want to be found and go home?
- Are you with other animals?
- Are there any people with you?

Some animals leave of their own accord and, for whatever reason, don't want to go home. Some animals might not be able to get home. They might be locked in a house or garden and can't get

# THE INTUITIVE EXPERIENCE

out, or there might be too many dogs not allowing them to walk past. As I found with one cat who told me she wanted to go home but there were too many dogs barking along the street and she didn't want to walk past them. Her experience had started off as an adventure. The other cats in her family had dared her to go and explore, as she had never gone out before.

Any form of stress, or an accident, can disorientate an animal, making it difficult for them to find their way home.

Animals are very psychic, so the first thing I advise someone to do when they phone me to track their animal is to picture their house in a white light and ask their animal to follow the white light home if they can. Also, place the animal in a bubble of white light for protection. Very often the animal will come back on its own.

## Tracking Shingi

Shingi is the most beautiful huge gray and white cat, who went missing on an Easter Friday. Shingi's mom phoned Carrie, a friend of mine, and asked her please to try to find him. Carrie asked her for a photo so that she could try to track him. The photos arrived late that night, so Carrie, with pen and paper in hand, took the photos of Shingi off to meditate and see what she could find.

Very quickly she connected with Shingi and asked him where he was and if he was okay.

He said, "Cold and dark, I'm hungry and scared."

She asked what his favorite food was so that she could verify with his mom that it was Shingi.

"Tuna, I love tuna," he answered.

# THE INTUITIVE EXPERIENCE

She asked him to elaborate where he was.

"At the recreation center near the squash courts." He then showed her the squash courts and a grid.

"Ask mommy to bring a tin of tuna, and some biscuits would be nice, too."

"Why did you run away?" Carrie asked.

"Thembi, the dog, dared me to as an adventure 'cos I never do anything daring. Ever."

When Carrie told Shingi's mom, she confirmed that he did indeed like tuna and there was a recreation center down the road. Wow, Carrie had indeed connected with Shingi.

The next morning, Carrie and Shingi's mom went to the recreation center on their quest to find Shingi. When they got there, they saw a security guard who told them he had seen a big fat gray and white cat near the squash courts. They left the tin of tuna with the guard and asked him if he saw Shingi again to give him the tuna and to phone them.

On Easter Sunday, Carrie connected with Shingi again. This time he showed her a small light green building which he was looking at. In picture form, he showed himself trying to hunt, there was a concrete wall to his right, with long grass. The grass that he was sitting on was shorter. He was trying to catch a lizard. Not much of a meal, he said.

Carrie sent Shingi white light from her heart to his heart. Animals will often follow the white light, but Shingi said he couldn't follow the light because there were too many dogs and he was scared.

# THE INTUITIVE EXPERIENCE

"I need help, come and fetch me."

The next day Carrie heard, "I've moved, I'm safer now: I've moved."

This confused Carrie and she started thinking, "What kind of communicator am I? I can't even find a cat." She started doubting herself. So she decided to phone me to see what I picked up. I later phoned them to say I had seen Shingi getting picked up by a little boy who had taken him home and closed him in his bedroom.

Shingi's mom felt better to know that he was safe but was getting desperate to have him home with her.

That night, there was a terrible storm with thunder and lightning, and everyone was panic-stricken worrying about Shingi out in the storm.

The following day Shingi's mom received a call, responding to the ads she had put up every where. Shingi had been found. A little boy had picked him up, taken him home and closed him in his room where he was happily sleeping on his bed.

Since then he hasn't left his mom's side and keeps prodding her face with his paw to make sure he is not dreaming. We all said a very big thank you to all our spirit guides for helping us get Shingi home safely.

## 🐾 Ginger and Gestalt

Gestalt is not a gift that only a few people possess, but rather a learned skill. The more you practise, the better you get. It is a technique of being able to get inside the animal and seeing out of their eyes. Once your animal is lost, there is so much stress that it makes it difficult to focus and do an effective Gestalt.

Ginger (the same ginger cat who used to live next door and came

# THE INTUITIVE EXPERIENCE

to live with me five years ago) is the most beautiful little ginger cat. He has big pale blue eyes, soft ginger fur, a big round belly that a Sumo wrestler would be proud of, a folded ear where he had been bitten by a dog before he joined my family, and short stubby legs. Ginger is a very wise cat with lots of attitude.

As he was walking past me in the lounge one night, I thought I would love to see what it would be like to be Ginger, so I used the Gestalt technique. Within seconds I could feel what it was like to be Ginger with fur on my face, stubby little legs, folded ear and big blue eyes. He was truly a magnificent little cat.

However, I now had to teach myself to disconnect from Ginger, because I found that every time I looked at him, I became him. He was very easy to connect with. After connecting with Ginger for the third night in a row, he walked past me and, lo and behold, if I hadn't seen it with my own eyes, I wouldn't have believed it. There was Ginger walking around with my face on his little body. I wondered whether he was trying to see what is was like being me.

### Riff-Raff's Gestalt

When I started practising the Gestalt technique, I decided to see where Riff-Raff was. I sat down on the bed and relaxed, and slowly visualized myself becoming him. While I concentrated on my breathing, I suddenly became aware of a lot of fur, all I could see in front of me was fur. I thought this was strange and decided to see where he was and what he was up to. I found him in the garden playing with Stacey, he had his nose in her neck, and she was one ball of fluff in front of his face.

### Orange

This story, as told by Ivan about the day he lost his beloved

# THE INTUITIVE EXPERIENCE

Orange, is a story with a happy ending.

It seemed like a good idea, to bring 'Orange' from France with me to South Africa. My work had brought me to Johannesburg and I had found a nice spacious apartment with a balcony overlooking a large internal garden. Although the apartment was on the third floor, Orange was 'bright' and I felt sure that he would be able to figure out how to negotiate the corridors and stairs in order to have unrestricted use of the garden.

I had adopted Orange five years ago, on the day I stopped the farmer next door from shooting him. I had heard loud gunshots and rushed to the window overlooking the farm just in time to see the farmer shoot the second of the three, six-month old kittens, who had been growing up on the farm. I yelled out that I was keeping the 'orange' one and not to kill him. The farmer shrugged his shoulders and Orange fled into hiding.

For sure he knew that I had 'saved his life'. He became incredibly affectionate towards me and for two years I was the only person to know of his existence. Nobody ever saw him because the moment anyone arrived at the house he vanished, only to magically reappear once the coast was clear. Gradually, he allowed me to introduce him to other people, as long as the introduction was 'formal' and ritualised. Slowly, his confidence in the human race grew.

All the arrangements to bring him to South Africa were made. Vaccinations, blood test, microchip, passport and finally my local vet assured me that all was in order. I purchased an expensive carrier and an even more expensive airline ticket and after a huge row with Air France who said Orange couldn't travel on the same flight as me (what they meant was he couldn't travel in the cabin with me) we were all set.

# THE INTUITIVE EXPERIENCE

On arriving in Johannesburg, I dashed to air cargo to pick up Orange. A wave of relief flooded over me when I saw him in his carrier, looking a bit startled but not much the worse for wear. We then had to wait for the South African state vet to "OK" the papers, which my French vet had promised me were in total order. They weren't.

The French vet should have had the papers countersigned by the French state vet, which is the legal requirement. The SA state vet wanted to send Orange back on the next plane to France and reluctantly agreed to keep him in quarantine, providing the required document arrived at the quarantine station within four days.

I won't go into the horrors of trying to get the papers signed and sent from France to South Africa (over a week-end) in order to meet the deadline. Finally, at the eleventh hour, the papers arrived and Orange, who I hadn't been allowed to see while he was in quarantine, was released.

Understandably, he was in a pretty disturbed state, not helped because I had to go out of town for a week to work and had to arrange for someone to come to the flat to feed him. I did several risky late night drives back to Johannesburg during the week to spend time with him but he wouldn't settle, and became highly agitated, specially between 3 and 4 every morning.

He spent many hours sitting on the balcony staring into the garden below, but when I tried to teach him how to negotiate the corridors and stairs he became paralysed and finally wouldn't leave the apartment. It was clear that he was bored, and longed to be in the garden, but couldn't find the way.

Then my friend, Carol, who lived in an apartment on the ground floor, with direct access to the garden, offered me the use of her

# THE INTUITIVE EXPERIENCE

apartment for a week while she was away. Orange and I moved in. He was in heaven, the longed-for garden was his! Within minutes he had figured it out, exploring every nook and cranny, disappearing and reappearing at frequent intervals. He reverted to the cat he had been in France – full of energy and fun, slept a full undisturbed night on my (borrowed) bed and was in and out of the apartment like a yo-yo. This happy state of affairs lasted for a week.

Then, one morning when I awoke and stretched out my hand to where he should have been...nothing, an empty space. My heart sank, and felt as though it had stopped beating.

Only those who have lost a pet know the agony of not knowing, or imagining, what has happened to them. The knotted feeling in the stomach, the uncontrollable bursting into tears, the feelings of overwhelming sadness and despair, swinging to feelings of possible hope, and back to despair. Add to this, the feelings of guilt at having brought him to South Africa. And now he was lost, maybe dead.

I scoured the area, printed hundreds of "lost" notices offering a reward, pasted them on every available lamp-post, tree, garage door, spoke with every security guard in the area, day shift, night shift. Cried out his name at midnight until every dog within the radius of a mile barked in response, went on every false lead stemming from the reward notice and, after ten days, felt exhausted, depressed and helpless.

Then something in my memory "clicked". I remembered picking up a magazine called Link Up and seeing an advert for "Animal Healing". I found the magazine with the ad and logged on to the website. "Animal tracking for lost animals." I was on the phone in a flash. A pleasant sounding woman answered (I later discovered this was Jenny Shone). She said her partner, Sandy, did the

# THE INTUITIVE EXPERIENCE

tracking and could I send her a photo via email. I said I couldn't as all my photos were in France. Then please send a detailed description instead...orange and white, beauty spot on the right cheek by the mouth.... . Sounds gorgeous, replied Sandy by email, give me a day or two to try to locate him...

Next day, in the street, my cell rang. Sandy: 'I've made contact with Orange.' As I have a problem with my hearing, Sandy said she would send me an email with what she had picked up from Orange. I hurried home. The information in the email was incredible. There is no way on God's earth she could have known the details she gave unless she had either broken into my apartment (the pale blue patterned sofa which Carol had lent me, the vase of flowers Orange loved, he would drive me mad walking over my computer keyboard in order to nibble at the foliage), or had a detailed conversation with either Orange or myself, and I had described only Orange to Sandy, not the apartment. I was stunned. Sandy said that Orange had gone searching for other cats (he had cat buddies on the farm in France) but was in good physical condition and wasn't hungry, but he was totally lost, alone and wanted to be 'home'.

My heart started beating again. Hope started to flood through my veins. The information was so precise, so accurate, that even the most hardened cynic would be hard put to explain 'how'. Other sessions followed, he was still in good physical condition, sheltering under something, perhaps wooden seating like in a sports stadium. Later, the image of water and long grass. He wanted to come 'home' but didn't know how, and although Sandy couldn't pinpoint where he was exactly she felt sure he wasn't far away. I was to visualize a white light in order to lead him home. But this was to no avail, and more days went by.

Finally, Sandy agreed to use a pendulum and a map in order to try and locate him precisely. I asked her, when she next contacted

# THE INTUITIVE EXPERIENCE

Orange (by this time I was totally convinced she was in contact with him), to ask him to 'show' himself and go up to someone he trusted. He was wearing a collar with an ID with his address and phone number, but unless he overcame his tendency to remain hidden and his desire to be 'formally' introduced, the ID would remain useless.

The next day the phone rang. Orange had been found, almost three weeks from when he had gone missing. He had been sheltering in a woman's roof, under the wooden eves, ten minutes walk away. He had been driving the woman mad, running around above her ceiling at night. She had had part of the roof blocked off by builders to try and curtail his activities and had called every animal organisation in Johannesburg to come and take him away. None would help and she was so distraught that she was about to call the exterminators to have him destroyed.

Then, the evening before I received the phone call, the woman and her gardener, Clifford, saw Orange for the first time, sitting on a table on the terrace. The next morning, Clifford saw one of the reward notices and copied down the phone number. I went to fetch Orange, there he was in perfect physical condition, (they found a lot of feathers in the roof!) and obviously delighted to have been 'found' by his human.

I immediately called Sandy, who was delighted at the news. Everyone was happy, I got my best friend back, Clifford got his reward, the woman got her peace and quiet, and Orange got to go back to France and settled in within minutes of arriving.

He is there now, being fed in my absence. I miss him so much that I am now taking telepathic communication lessons from Jenny Shone to stay in touch with him...long distance...it's the least I can do.

# THE INTUITIVE EXPERIENCE

Although Sandy finally didn't need to use her pendulum to locate him (Clifford beat her to it), she gave me the hope and belief that he was alive and the determination that I would find him, however long it might take.

Orange has since been sent back to France where he asked to go and be with his friends. Ivan regularly visits him in France. He is once more a very happy kitty.

# EXERCISE 3
## Feeling emotions

Here you can either visit a zoo, a friend or stay at home with your animals.

Whether an animal approaches you or you approach an animal, always take note of your first thoughts and feelings.

Do you feel anger, joy, insecurity, pain, excitement or acceptance?

Do this as often as you can, feeling the emotions of the animals around you.

*Jenny connecting to Gillie, a young zebra*

# CHAPTER Five

## THE CONTACT EXPERIENCE DURING GRIEF
### Heavenly voices

*We each have our own special purpose.
We all have a lot to contribute;
you must listen to all of us.*

***Rebecca***

# THE CONTACT EXPERIENCE DURING GRIEF

### 🐾 Meeting the souls of all my animals on the other side

I had had a lovely deep meditation and was feeling very relaxed. All I could hear was the sound of the birds in the garden and the whistling of the wind through the trees. It felt as if I was totally connected to the entire universe. Just as I was about to come out of my meditation, I suddenly got a feeling to ask my guides to help me connect to the souls of every animal I had ever spent time with, in all my past lives up until now.

Suddenly I was in a big field of autumn colours, all different shades. This field suddenly became full of what looked like fireflies. Everywhere I looked, there were little sparks of light, hundreds and thousands of bright sparkling lights. I was amazed as I looked at this field full of tiny little sparkling lights. It only lasted a few seconds but I knew these were the souls of all the animals I had ever shared my life with.

I had just connected with all my beloved pets and at that moment I realized they would always be with me. I would never be alone again.

This next story I would like to share with you is about Gizmo, the most beautiful little dog. I feel honored to have been the one asked to communicate with her.

### 🐾 Gizmo

The reading I did on Gizmo was one of the most amazing readings I have ever experienced with an animal that has crossed over. I really enjoyed connecting to Gizmo and receiving her messages.

I was contacted by Theresa, Gizmo's mom, who asked me if I could contact Gizmo as she missed her very much and wanted to know if she was all right, where she was and who she was with.

# THE CONTACT EXPERIENCE DURING GRIEF

Before I even got the chance to connect to Gizmo, she connected with me. I found her very talkative and playful, and she had such character. I had a lot of fun chatting to her.

This is what she said to Theresa through me.

"I am very happy here. You helped me when I needed it and I thank you for that. The fact that you put aside your own feelings to help me when you didn't want to; it was a hard decision and proves your unconditional love for me. It was my time, I was ready to go, don't feel guilty.
"Although I'm on the other side, I'm still with you and will be as long as you need me.

"Don't feel sad, we had a good life together. I can feel your thoughts. When you feel sad, I feel sad; and when you feel happy, I feel happy.

I still play with the yellow ball, and I still sleep at your feet on the bed with you at night.

"The only thing that's gone is my physical body, my soul is still the same as it was when I lived with you on earth.

"If you stop and look back, there were many lessons I taught you. Go forward, use these lessons, make them part of your daily life. I've got so much love in my heart sometimes I want to burst. Use my love to heal your pain.

"I will be back one day, you will know me. But for now, give your love to someone else who needs it. Watch for signs and listen. I will always be there and you will know it's me.

"There are lots of dogs here with me and some cats (I love chasing them), and there are people, too. I'm having so much fun. Every one here looks up to me. I'm very important.

# THE CONTACT EXPERIENCE DURING GRIEF

"I want you to connect with me so I can pass on my wisdom to you and guide you through your life. Thank you for being so special and looking after me so well while I was on earth with you.

"Don't forget: I love you always."

These were words that came from such a wise little dog.

When you connect to any animal, you are connecting to the soul of the animal. When an animal dies, only the physical body dies, the soul lives on in another dimension and never dies. For this reason, it is just as easy to connect to a living animal as it is to an animal on the other side.

Every animal, human and even plants have an energy field called the aura. This energy field has a vibration, and everyone vibrates at different speeds. On earth, the aura is very dense and vibrates at a slow speed, which can be improved through meditation. People and animals on the other side vibrate at a very fast speed, making it an effort to appear to us here on earth. In order for us to see or hear them, they need to slow down their vibration, and we need to speed up our vibration so we can meet them half way.

If an animal on the other side has a message for us, they will find a way to get through to us and let us know. However, we must remember the soul has come to earth to have certain experiences. Once an animal has crossed over, it is not fair for us to keep calling them back. They now have their soul experience to go through before deciding whether or not to come back in another body for more experiences here on earth. They may decide to stay on the other side to further their soul experience, or even to have an experience of being a guide. Pining for them keeps them here with us and does not allow them to move on.

Mourning for your lost pet is completely normal. It is something everyone must go through. However, when the time comes, you

# THE CONTACT EXPERIENCE DURING GRIEF

need to release them and let them go.

All this does not mean that when they die, they are gone forever. Once you have let go and allowed them to move on, they will still, although having their soul experience, come to visit you and be there when you need them.

I have communicated with many animals (my own included) that have always said they were very happy and healthy, and have thanked their people for helping them to cross over at the end. They say it is the ultimate act of love for someone to put aside their own feelings and help them cross over when they really need it. We call it murder, they call it love. There is no regret, no anger, only a deep sense of love.

Animals don't have the same concept of death that we do. They have a far bigger understanding and acceptance, and we could learn a lot from our animal friends. The only thing they ask is that their human who loves them on earth is with them at the end. They don't want us to feel guilty, just happy that we could be there for them at the end.

When you connect to an animal through a photograph, it is not the photograph that you are connecting with but the soul of the animal pictured in the photograph. The photograph acts as a visual aid, and this makes it easier to connect to its soul. When you communicate with an animal, it is the soul of that animal you are connecting with.

### 🐾 Christopher Robin

Many years before I started doing this work, I went to see a psychic woman for a reading. She asked me if I would like to experience a past life as well as the other side. I said yes, so she started leading me through a meditation.

# THE CONTACT EXPERIENCE DURING GRIEF

I was totally relaxed, when suddenly I found myself walking in the most beautiful countryside. In the distance, I saw a horse galloping towards me. As it got up to me, I saw it was Christopher Robin, a beautiful chestnut gelding that had lived with me for twenty five of his years. A few years earlier, he had collapsed and it was necessary to have him put down since his heart had failed. It was a very upsetting time for me as he was my friend and I loved him.

I was amazed. I wasn't trying to contact him at this time, yet here he was with a message for me. He said he was waiting for me, he loved me and when it was my time to go, he would be there for me, he would meet me when I crossed over. I stood there with my head buried in his neck, hugging him. We stood like that for quite a while before I eventually left and was brought out of my meditation.

I was thrilled to know Christopher Robin was there for me, waiting for me and loving me. I miss him and love him and thank him for all the fun he brought to me while we shared a life together.

 Tess

There are many different ways animals, or even people for that matter, get messages to you from the other side. They could send a message through a song on the radio, a billboard that keeps popping up, advertisements on the sides of buses. Have you ever wondered why you suddenly start thinking of someone you lost and haven't thought of for years? They could be putting the thoughts in your head, hoping you will connect with them.

Tess was a beautiful Rhodesian Ridgeback who came to live with me after her people immigrated many, many years ago. She was seven months old at the time. She was a very sensitive, intelligent and loyal dog. She was the most responsive dog I've ever trained. She loved her dog training and did well in agility. Tess lived with

# THE CONTACT EXPERIENCE DURING GRIEF

me for about twelve years until I lost her to organ failure.

I had thought about Tess a lot, but for some reason didn't communicate with her when she passed to the other side.

About three years after the loss of Tess, I went to see a movie with my husband, Alan. During the advertisements, a music video started playing. It was a terrible, boring song and I wanted it to finish. The movie ended and we went home.

A few weeks later, I got into the car and the same song started playing on the radio. I thought, what a boring song. On my short trip to the shops and back that same song played three times. For the next month, every time I got into the car the same song would come on within a few minutes and play at least three times before I got home.

Alan commented, "Why does this song play every time you get into the car?" Yet, when he got in on his own, it didn't play. I jokingly said, "It must be someone trying to get a message to me."

I started to listen to the song with a different outlook. What were the words? I discovered the words that stuck out over and over again were, "I love you, I love you, I love you." I started to wonder, was it my dad? No, it wasn't. Was it Smurfie? No. Was it Maggie? No. Then I suddenly thought, "I know who it is. It's Tess."

I acknowledged it was Tess, I told her I loved her and would contact her because I missed her very much. The minute I acknowledged Tess and started to communicate with her more often, the song stopped and I haven't heard it since. That was six months ago.

The reason she chose that song was because she knew it would

# THE CONTACT EXPERIENCE DURING GRIEF

make me listen, as it was a song I didn't like. Every time I heard it, I wondered, why this song?

Thank you, Tess, for opening my eyes.

### 🐾 Maggie

Maggie was one of the most amazing dogs. She was a black and white Bull Terrier/Labrador cross. Some of you may have seen her face on the Animal Healing Center logo together with a zebra.

Maggie was a healer and teacher, she taught many people how to communicate with animals. Maggie always felt humans needed to communicate with each other before learning to communicate with animals. She was concerned about the human race. I learnt a lot from Maggie in the seventeen years she lived with me before crossing over.

Maggie had many tactics to get her way. If she wanted to sleep where someone else was already sleeping, she would tiptoe to the door and give one loud yap. The other dogs would all jump up to see who was coming and Maggie would quickly run to the bed, jump in and fall asleep. The other dogs would then have to find somewhere else to sleep. It worked every time.

It was a very sad day for all of us when Maggie crossed over. I'm just very happy to say that, today, Maggie has appointed herself as one of my guides. She comes on all my workshops to help the people I teach and the animals concerned to communicate with each other.

### 🐾 Nandi

Nandi was a bay mare whom I had from birth. Her mother, Bess, was my first horse. Nandi was only eight months old when Bess

# THE CONTACT EXPERIENCE DURING GRIEF

died of African Horse Sickness. We were all devastated. I became Nandi's mother and looked after her.

When she was three years old, I started her training. We had a lot of fun together and developed a bond that no one could break. I introduced her to her first saddle and started riding her when she was four. We would go on outrides and have lots of fun. Nandi, together with Christopher Robin, became my leading horses in the riding school I started. They taught many children to ride.

Nandi was the most talkative horse I have ever known. Every time she saw me in the distance, she would call to me. If I happened to walk past the bedroom window, and she saw me from her stable, she would shout and wake up all the other horses. I loved Nandi, she was great.

Quite a few years later, Nandi had a foal who I named Topaz. He was a thriving young boy with a lot of attitude. Eventually I retired her, but as she would get bored, I would take her out for a twenty-minute walk twice a week.

One morning before the sun came up, Nandi developed a twist in her bowl. The vet said we should try to get her to hospital to operate. I knew she wouldn't make it and decided to let her go. She was in a lot of pain and would never have made the trip to the hospital. I wanted her to go peacefully where she had been so happy for so many years and not in some cold hospital. The vet agreed and I sat with her head on my lap while she quickly drifted off. She was at peace and her pain was gone. Now, all that was left for all of us, who shared her life, was to come to terms with our loss. Nandi was thirty-seven years old when she left us.

## 🐾 Topaz

When Nandi passed over, Topaz, Nandi's son, looked fine. He had

# THE CONTACT EXPERIENCE DURING GRIEF

all the other horses with him, but I hadn't realized how much the loss of Nandi had affected him. He was always full of mischief and every time there was any mishap, Topaz was in the middle of it. Every show he went to, he would see who he could buck off first. Once someone had been bucked off, he would settle down and win prizes.

Topaz took it upon himself to be the keeper of the miniature horses. He would herd them around, and discipline them from time to time.

Thomas would ride Topaz, I would ride Rebecca and off we would go for long walks in the beautiful countryside. Thomas also had a great bond with Topaz.

It was exactly one month after Nandi's passing that Topaz developed exactly the same problem. He was eighteen years old and quite a lot stronger than Nandi, so we decided to operate on him at home. With horses, this is always very risky but we didn't have a choice.

The operation seemed to go well but we had to wait and see. Topaz started to improve and even walked around grazing, but later that afternoon he started to deteriorate. I called in the vet.

Topaz had decided it was his time to go. He didn't want to be here without Nandi. Together with the vet, Thomas and I sat with Topaz while he slipped away to be with Nandi.

I couldn't believe it. I had lost three of my best friends in the space of one month. I wondered what would I learn from this experience?

## Smokey

Alba had a cat called Smokey. He was a young, beautiful Siamese cat and everyone loved him dearly. His sudden passing a couple of years ago was very traumatic for Alba. She lost three of her

# THE CONTACT EXPERIENCE DURING GRIEF

animals in two weeks, of which Smokey was the second. One day while I was visiting her she asked me if I could connect with Smokey and see how he was.

A few days later I sat down to connect with Smokey.
I connected almost immediately. Smokey was looking very regal. He said he was a very old soul. This meant he had had many lives on earth and was quite advanced spiritually.

He had a very strong energy around him. There were lots of cats and dogs with him. He said he was in charge of all of them. He was their superior.

He had a special friend with him, a black shaggy dog.

Smokey was there to teach other animals. While I was chatting to Smokey I kept on getting flashes of female energy. He was a very sensitive cat. He told me he was a rescue worker helping other animals cross over. He said they needed him so he had to go and that is why he left life early. I asked him how he passed away. He replied that it is not important.

He carried on speaking. " I love my mom. I'm very proud of her. I feel so light, so happy. I love it here."

One year later and I am connecting to Smokes, as he likes to be called, to get some more information from him.

I asked him to tell me more about his job as a rescue worker.
Here is his story.

"It's a very rewarding job and I love it. I help all animals that arrive here. Some arrive after passing naturally and I welcome them home, helping them to readjust to their lives over here. Those that pass after a long illness or a sudden accident I take to a safe place - almost like a hospital where they get put into a sleep state and we do healing on them helping them recover. This is the reason why some animals don't communicate straight after

# THE CONTACT EXPERIENCE DURING GRIEF

passing. It sometimes takes a few weeks or even months.

"Others arrive here and get straight into work the way I did. Not all animals know what their jobs are until they get here. Not all animals are rescue workers. Some are companion animals, some are teachers. Just like humans on earth, they all have different jobs. There are many jobs for the animals here. What I am doing at the moment is working with humans and animals.

"I have moved on, but only to a different duty - not a different place. I still come to visit my earth family at times.

"I will be going to have another earth experience, but not just yet. At the moment I'm just too busy and loving it.

"Losing an animal friend, although devastating, is part of the human experience. It is important to your spiritual growth. Even if the human knows that their animal is a rescue worker, it is still devastating.

"My message to all humans is:
Communicate with and get in touch with the soul of your animal friend before they pass. Feel the love and celebrate their lives on earth and over here. Do not feel anger or pain at their passing. Be happy and know, we are happy."

# MEDITATION
## 🐾 Connecting to pets on the other side 🐾

For this meditation you can use a photograph of the pet that has passed on. Make sure you are sitting where it is very quiet and you feel safe and relaxed.

Place the photograph on your lap, or just keep the memory of the animal in your mind. See yourself walking to the bottom of a garden filled with beautiful flowers. There is a glass lift at the bottom of this garden. You invite your guide to join you on this journey. Together, you and your guide climb into the glass lift, the doors close and the lift starts to climb. As you go higher and higher, you look out and see a field of beautiful red flowers. All around, as far as you can see, there are red flowers. The flowers are magnificent, they give you a feeling of being totally grounded and connected to the earth. You are safe and happy. As you climb higher, you become aware of a field of orange flowers. Feel the essence of these orange flowers. Feel your creativity bloom as you move through these orange flowers. Now, as you go higher, the flowers you see are all a beautiful yellow; they are glowing with this beautiful yellow colour. Become aware of all your emotions, your excitement, as you move through these yellow flowers. You now look out of your glass lift and see all around you exquisite green flowers. As you admire these green flowers, you feel the love radiating from these green flowers and from your heart. You feel almost ready to explode with love. You now find yourself moving through a field of brilliant blue flowers. As you move through these flowers, smell their fragrance, feel their essence. Feel your centre of communication opening. As you keep going higher and higher, you now see indigo flowers everywhere. Beautiful, vibrant, indigo flowers. Feel your psychic abilities opening up, your intuition becoming sharp and aware. You marvel at the

*continued...*

beauty around you. You keep climbing, everything around you is a beautiful violet colour, all the flowers are violet. You feel yourself connect to your higher self, God, Goddess, your guides and angels.

As you look around now, you see pure, bright, brilliant white. It leaves you feeling safe, secure and excited. As you climb higher and higher through this brilliant white, you start to hear the faint sound of angels singing. Finally, the lift stops, and you and your guide get out. You can't believe the beauty that you see all around you. In the distance you see a gold ball of light rolling towards you. The gold ball stops right in front of you. Out of the center of this gold ball of light steps your special friend that you have come to meet here today. You hug and kiss each other until you cannot hug and kiss anymore, then you walk off together side by side. Now spend a few moments with your special friend, catching up on news and asking any questions you feel you would like your friend to answer. It is now time for you to leave. You hug and kiss your special friend. Thank him or her for coming to meet you here today. As you climb back into the glass lift with your guide, you feel happy to know that any time you want to visit your special friend, you can. For he or she is not gone, but just in another dimension. The lift starts to descend, down past the white, into the violet, through the indigo. Lower and lower, until you reach the blue. As you go lower, feel the love, become aware of the green. Move down to the yellow and become aware that you are getting closer and closer to the ground.

You feel totally safe and relaxed. You now look around and see orange. Start to be aware of your breathing. Now you see the red, and you feel completely grounded as you wiggle your fingers and toes. Become aware of the beating of your heart. Take a deep breath in, hold it for a moment and, as you exhale, feel yourself relax. Listen to the sounds around you and thank your guide for joining you on this journey.

Then, in your own time, open your eyes.

# EXERCISE 4
## 🐾 Getting permission 🐾

It is very important not to take our animal friends for granted.

This exercise is best done with an animal that you are not familiar with.

Show this animal the utmost respect and explain that you are practising your communication skills, and you would like to communicate with it.

Ask this animal for permission to touch it, and feel if you get a yes or no answer.

Then touch it and see what response you get.

Thank the animal for allowing you to touch it.

Practise this with as many animals as possible.

# The Rainbow Bridge

There is a bridge connecting Heaven and Earth.
It is called the Rainbow Bridge because of its many colours.

Just this side of the Rainbow Bridge,
there is a land of meadows,
hills and valleys with lush green grass.

When a beloved pet dies, the pet goes to this place.
There is always food and water, and warm spring weather.

Those animals who are old and frail are young again.
Those who have been maimed are made whole again.
They play all day with each other.
But there is only one thing missing.
They are not with their special person
who loved them on earth.
So, each day, they run and play until the day comes
when one suddenly stops playing and looks up.

The nose is twitching, the ears are up, the eyes are staring,
and this one runs from the group.
You have been seen,
and when you and your special friend meet,
you take him or her in your arms and embrace.

Your face is kissed again and again and again
and you look once more into the eyes of your trusting pet.

Then you cross the Rainbow Bridge together,
never again to be separated.

*Paul C Dahm*

# CHAPTER Six

## THE SOUL EXPERIENCE -
### death and rebirth

*When you awake in the morning,
feel the sun rise from the centre of your heart.
When you are busy with your day,
feel the wind push you on your way.
When you sleep at night,
feel the moon deep within your soul.
Relax, for you are alive.
Be thankful, be happy.*

**Ginger**

# THE SOUL EXPERIENCE

The soul is energy in its purest form. It is timeless, meaning it never dies.

Everything from human, animal and plant is made up of energy.

Before a soul (animal or human) decides to get born into a physical form to have a human or animal experience, it makes a decision and chooses the body it will be born into.

The soul has only one life experience, but many bodies. Some people think of this as reincarnation.

As I mentioned in Chapter Five, the soul never dies but lives on. Only the physical body dies. The soul moves on to another body, giving it the opportunity to have more life experiences. Most of us have lived with our pets in many different lives.

As the soul has chosen to incarnate into this life and have certain experiences in it, these life experiences are the most important to the soul at this time. Once the soul has crossed over, it is now in another dimension and is having a soul experience. We should respect this and allow the soul to move on. It doesn't mean that you can never connect to a pet once it has moved on – you can always allow it to come through with a message from time to time. However, by holding on to it once it has crossed over, you are preventing it from moving on.

Most people think a soul spends its time floating around enjoying the beauty and doing nothing, except waiting for us to make contact with it.

Not so, a soul has a purpose, a life, a job and friends. It is just in another dimension where there is still work to be done. Some souls act as rescue workers helping new souls to adjust to being back in that dimension. This is especially the case for those who have

# THE SOUL EXPERIENCE

passed over in a violent or traumatic way, such as an accident, or after a lengthy illness.

If you feel that your animal friend is ready to go but is holding on after a long illness, you can give them permission to go. Tell them that you will grieve for them and miss them, but you will be fine. Sometimes they are hanging on for you, because they are worried you won't cope and they don't want to let you down. Once you have assured them that it is okay, they can then peacefully leave, and move from a state of pain into a state of shear bliss. You have given them the biggest gift of all.

I recently was asked to connect to a little Jack Russell dog that had been hit by a car. The vet had said he would be fine, but four days later, to everyone's surprise and distress, he crossed over. On the same day that he crossed over, his people contacted me. I explained to them that I preferred to wait a couple of weeks before contacting an animal that had crossed over, because they usually needed some time to adjust.

They were so distraught and just wanted to know if he was okay. I thought I would try to connect to him the following day. When I connected with him, he came through loud and clear. He was fine, and was with human and animal friends he hadn't seen for a long time. He was too busy having fun playing with his friends to chat now. "Could I come back another time?"

I relayed this to his people. Although they missed their little dog terribly, they felt so much better knowing he was fine.

Before the animal or person is born, the first thing that arrives is the energy (aura). Once all the different layers of energy are in place, the soul enters.

As difficult as this is to understand, the soul is here to have life

# THE SOUL EXPERIENCE

experiences. The more difficult the experience, the faster and bigger the spiritual growth will be.

When you do a healing on a sick animal, it doesn't always work the way you expect or even want it to. Sometimes the animal still crosses over. This is not because the healing didn't work, but rather because, for whatever reason, it is the animal's soul choice to cross over at that time. On some level, the soul needs to go through a certain experience and, as difficult as it is, you need to respect this. However, what the healing has done, is to make it easier for the animal or human to cross over.

There are many people who are against euthanasia for various reasons. Some people feel it is the same as murder. It is really the ultimate, selfless act of love, helping the animals when they need it the most. All the animals ask is that their human, who shared so much love with them through their life, will be there with them at the end.

Once you can effectively communicate with your animal friend, you can ask them if they want to do it alone, or if they want help at the end. You can ask them for a sign to show you that they are ready and want your help.

There is nothing wrong with giving them a tranquilizer to calm them down, or even some Rescue Remedy®, which you can get from your chemist. Sometimes gently massaging their paws or ears will also have a calming effect.

Many people are so wrapped up in their own loss and despair that they forget that the other animals that are left behind are also grieving the loss of their friend.

This happened to me when I lost my beloved Smurfie. I went into a total decline and didn't notice the pain all the other animals were

# THE SOUL EXPERIENCE

going through.

Finally, after four long months, I realized what was happening and came to my senses. I shared the loss with my other animal companions and helped them to grieve. I took them out to where we had buried Smurfie and we sat together, chatted and remembered our friend with love, not pain. Later that day, we all went indoors where I lit a special candle and allowed all the animals to say good-bye.

I find that what also helps a great deal is to write a letter to the animal that has crossed over. Write all your feelings down on paper, the love you feel, the anger at the loss, the good times you shared, whatever comes to your mind, then burn it in your own little ceremony and let your animal friend go.

The animal that you have lost feels all your pain, and joy. They need to know that you will be okay.

# THE SOUL EXPERIENCE

This is a poem my sister, Linda, and her husband, Michael, as well as their son, Alexander, and daughter, Nicola, received from their vet after the loss of their beloved dog, Jonty.

## God Bless Our Pets

They say memories are golden,
Well, maybe that is true.
I never wanted memories,
I only wanted you.
A million times I needed you,
A million times I cried.
If love alone could have saved you,
You never would have died.
In life I loved you dearly,
In death I love you still.
In my heart you hold a place,
No one could ever fill.
If tears could build a stairway,
And heartache make a lane.
I'd walk the path to heaven,
And bring you back again.
Our family chain is broken,
And nothing seems the same.
But God calls us one by one,
The chain will link again.

*Anna Charlston*

# EXERCISE 5
## 🐾 Communicating 🐾

Find a comfortable safe place to sit with your animal friend.

Have a pen and paper ready.

Start by quieting your mind.

Connect with the green light from your heart to the animal's heart. Open your heart and send all the love you can.

Pick up any emotions going on inside this animal.

Now ask permission to chat to your animal friend and feel the answer.

If you get a "No" then try again later.
If you get a "Yes" then proceed.

Start with simple questions.

Write down the first thing that comes to your mind no matter how silly it may seem to you.

- What is your favorite food?
- Who is your favorite person?
- What is your favorite toy?
- What is your biggest fear?
- Who is your best friend?

When you have finished, thank your animal friend for chatting to you.

Reward him or her for helping you to practise.

# THE ANIMALS BEHIND THE
## *Voices*

*Meet all the animals
who have shared their
inspirational messages and stories.*

**Candy -**
*who was the reincarnation of Smurfie*

**Smurfie -**
*my very best friend and angel dog*

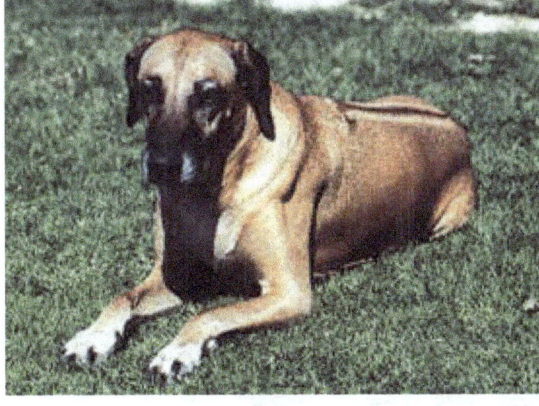

**Tess -**
*taught me to listen to messages coming from the other side*

**Maggie -**
*still helps and motivates me to this very day from the other side*

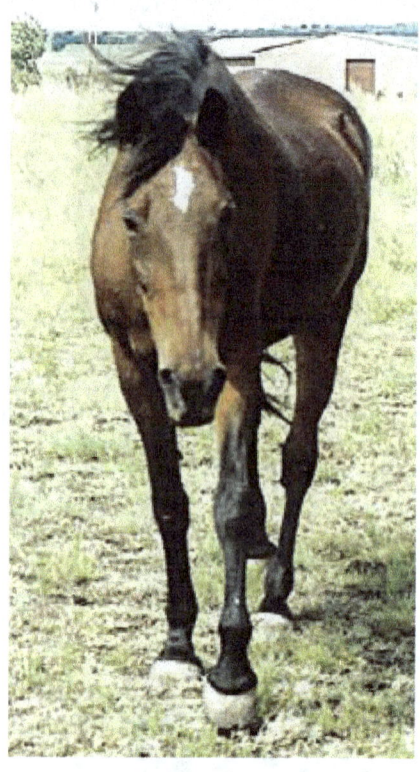

*A friend spends a moment with* ***Topaz***

***Nandy -***
*was the first horse I bred and was 37 years old when this photo was taken*

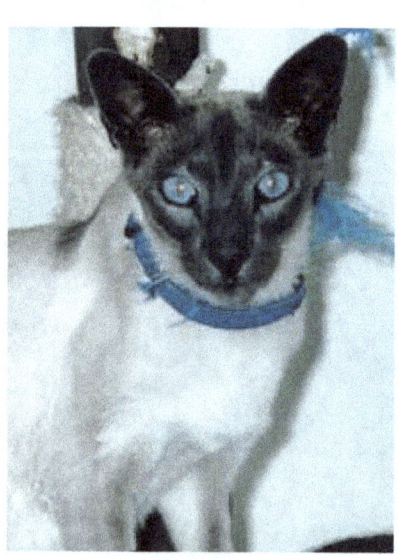

***Gizmo***

***Smokey -***
*now an animal rescue worker on the other side*

*Frodo*

*Stacey*

*Ballinore*

*Gabriel*

*Mona Lisa*

*Daisy*

*Snoopy*

*Ginger*

*Jenny and Riff-Raff*

*Some of the lions Jenny communicates with*

*Sandy and Jenny with **Edith** the Owl*

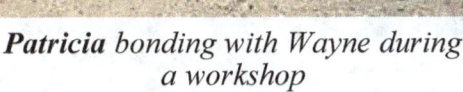

***Patricia** bonding with Wayne during a workshop*

*Jenny and Sandy at Monte Casino Bird Park*

**Merry** and **Pippin**

**Bella**

***Vista** was the guest speaker at this workshop*

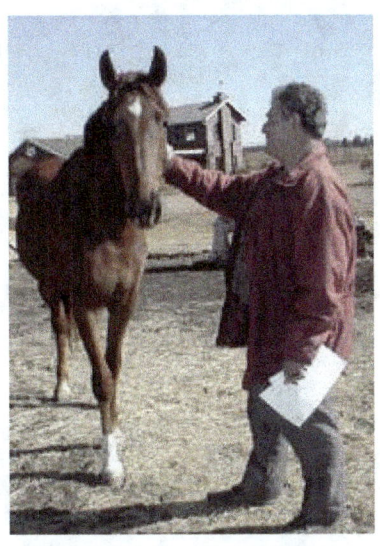

**Red** *during a workshop*

**Savannah** *balancing to grab her stick from the pool*

**Archie**

**Lucy,** *the day she arrived, with Alexander, Jenny's nephew*

**Penny**

*Jenny and **Tootsie** sharing a special moment*

***Isis** and **Ginger***

*Thomas, Kate and their daughter Jenny at the entrance to The Animal Healing Centre*

***Mishka** enjoying her garden*

***Rebecca** with Vernon and Barry*

# CHAPTER Seven

## TALK TO THE WILD SIDE -
### and voices from the ocean

*The world is there for all of us to share.
We are all a part of the same Universe
and we need to realize this.
For everyone
– human, animal and plant –
is here to make a difference.
We are all important in the bigger scheme of things.
The sooner we realize
that we are all part of the same oneness,
the sooner we will feel the love
of our fellow beings and
that of the Universe we live in.*

***Tootsie***

# TALK TO THE WILD SIDE

Communicating with domestic animals is a terrific amount of fun and you can achieve great results. It is just as rewarding and just as easy to communicate with wild animals.

However, people always say to me, "If I am an animal communicator and can talk to all animals, why can't I get out of the car in the game reserve and walk with the lions or elephants?"

The answer is simple. Even though you can communicate with a wild animal, you need to realize it is still a wild animal and has its wild instincts. You need to respect it for who and what it is. Being a communicator does not give you the right to demand of any animal a certain type of behavior.

If you ask a child not to eat a certain sweet because it is bad for her, does she always listen? If you ask your dog not to bite the cat, chances are when you are not looking, he will bite the cat. So if you ask a lion not to eat you, are you willing to take the chance that he will listen to you?

The key here is respect. Whether it is a dog, a cat, a horse, another human or any breed of wild animal, we all need to respect each other, and be respected in return. Never forget that we are all souls having certain experiences. We all have a right to live the life we have chosen.

## 🐾 Dad and the kudu

Many years ago, while I was a child, my parents decided to take my sisters and me to the local zoo to see the animals. As we walked along a cobbled path, we passed a number of different animals in camps. There were some zebra, a few different breeds of buck, some ostriches and some type of lama.

It was a hot sunny day and there was a slight breeze in the air. Up ahead, we could see a family with three children eating ice creams. But they weren't keeping the ice creams for themselves. They were sharing their ice creams with a very elegant looking kudu. We watched in amusement, as the kudu seemed to be enjoying the ice cream. As the family moved away, we moved in closer to look at this magnificent kudu.

My dad walked up to the kudu and started to explain to us, "If you want to make friends with a wild animal, you need to blow into its nose, like this."

He rounded his mouth ready to blow into the kudu's nose. But alas, at precisely the same time that he blew, the kudu, with his ice cream-coated slimy tongue, positioned a well-aimed blow. Bingo! With one fell swoop, he put his tongue right into Dad's mouth and down his throat.

Spit, spit, spit all over the place my dad went, while my very entertained mom rolled around on the ground, "Ha, ha, ha, ha, ha, ha, ha and ha!"

My dad never blew on any noses after that.

To this day, we still laugh when we remember my dad bonding with his kudu friend.

### 🐾 The crested barbets

These events happened many years ago, before we were aware of animal communication.

I was a little girl living in a house in Randburg. At the bottom of our large well-wooded garden and between the firtrees, there nestled a little thatched hut. One of the firtrees sent a branch across the hut's thatched roof causing damage. So it was decided to remove the tree.

The gardener was sent with chopper and saw to chop down the tree.

All day long, two crested barbets pecked at the front door of the main house at the other end of the property. My mother explained to them about the danger of the cats and that they should be careful. They took no notice and continued with renewed zeal to peck at the door.

When my father came home that evening, we all went down to inspect the fallen tree. With horror, we noticed a little hollow in the stump, lined with feathers.

"We have cut the barbets' nest in half," my mother shrieked, "we will have to mend it at once."

We lifted the section with the hole and fitted it to the base, restoring the nest. A tile over the top would keep it waterproof. That done, we walked back up to the front door and stood in the driveway under the jacaranda trees and my mother called out, "Barbets, we have fixed your nest, you can go back now." And they did.

We never heard them tap on the front door again.

## London Zoo

Just after I finished my communication course with Amelia on the Isle of Man, I decided to visit the London Zoo so that I could chat to some of the animals before flying home to South Africa.
The first animal I encountered was a very beautiful hawk.

# TALK TO THE WILD SIDE

I asked him, "How long have you lived here?"

He answered, "Five years."

"What kind of a bird are you?"

Sounding annoyed he replied, "What do I look like?"

"Do you like talking to me?"

"I always like to chat, it passes the time."

I then asked him if I could draw a picture of him.
"Of course you can. "

Animals are very vain and love you to compliment them, either by drawing them or just admiring them. I did a little sketch of him and showed him. He was happy. I thanked him again for the chat and went on my way.

As I came around the corner, there was a cage full of very playful monkeys. As I walked up to them, I said, "Hello, monkeys, can I chat to you?"

"Can't you see we're having too much fun to be bothered chatting to you? Come back later."

While I was busy being snubbed by the monkeys, in the distance I saw a crowd gathering and thought, "Let me go and see what is going on over there."

I pushed my way through the crowd and there in front of me was a very bored but exquisite camel. I quickly separated myself from the crowd and went to stand behind and off to the right side of the camel. He looked so bored.

# TALK TO THE WILD SIDE

"Hello, I'm an animal communicator from South Africa. If you can hear me and want to chat, I'm standing behind you and off to your right side. Please turn around, look at me and blink your left eye."

I wasn't asking for much!

A few seconds later, he turned around, looked at me and blinked his left eye. From that moment on, he didn't take his eyes off me for the duration of our conversation, which lasted about fifteen minutes.

I asked him, "What do you think of all these people?"

He answered, "Very entertaining…"

"How do you feel?"

"I'm a bit bored."

I asked, "Have you got any questions for me?"

"Give me some time and I'll think of some."

After chatting for a while, I asked him to wink at me when he was ready for me to go. He did, and I thanked him for the conversation. I left, and he turned around and started eating the turf that was in front of him. I was feeling a sense of awe at the privilege of being able to talk to these wonderful animals.

The next animal I encountered was a magnificent brown bear. As I approached him, I got a terrible sense of sadness. I sat with him for a while and just sent lots of love. I explained to him that although his situation was not perfect, he was protected from poachers, and he had food and shelter. I told him that he was providing us with a great opportunity to learn from him and to admire his magnificence. As I was sitting there, I could feel his sadness slowly lift. He would be okay.

# TALK TO THE WILD SIDE

Animals are very sensitive to our feelings and thoughts. They see the meaning behind our thoughts. If you walk into a zoo and see an animal in a cage or an enclosure and think, "Shame, poor animal," they will pick this up and immediately be dissatisfied. Most of them are reasonably happy since they don't know any other way of living. By appreciating them and telling them how lucky they are to have such a beautiful home with people to look after them, they will immediately feel secure, happy and content.

With all animals, you need to focus on the positive rather than dwelling on the negative.

## 🐾 Bird with a headache

Back at home, the early morning sun was coming up over the horizon, the air was crisp and clean, and I looked out over the paddocks to see the hills clearly in the distance. I was thankful to be living in this beautiful countryside with all my animal friends.

Then I suddenly heard this little voice, "Heal me, heal me, heal me."

I wondered where was this voice coming from. I looked up. Right in front of me, sitting on the branch of a tree was a little bird. I said, "Are you talking to me, little bird?"

"Yes, heal me."

While I sat looking at the little bird I noticed he had the biggest greenest aura I had ever seen on a bird, but around his head there was grayness.

"Do you have a headache?"

# TALK TO THE WILD SIDE

He replied, "Yes, heal me."

So I sat with my hands facing the little bird and beamed Reiki energy at him. As I was working, he slowly turned around for me to do the other side. Eventually, all the grayness was gone, and all that was left was a beautiful clear green aura.

He said, "Thank you," and flew off.

I was left sitting there, amazed at this beautiful little bird that had come to me for healing. At that moment, I was thankful for the ability to be able to communicate with, and heal, animals that needed it.

### 🐾 Talking to lions

While I was doing research for this book, I decided to build a relationship with some lions that lived near to where I stay.

Animals are very much like humans in that if you want to get deeper information from them, you need them to get to know you. You need to visit them regularly to build up a relationship.

If you meet someone for the first time, you tend to talk to him or her on a polite basis. Once they get to know you, they will be far more ready to give you deeper and more personal information.

I decided to do this with the lions. I would visit them regularly, let them get to know me, and eventually get some deeper message from them.

The depth of the message you get also depends on the depth of the listening you do. If you listen on a shallow level, you will get a shallow message, but if you listen on a deep level, you will get a deeper message.

# TALK TO THE WILD SIDE

With this in mind, I set out with my sister, Linda, and a friend, Carrie, to meet the lions one Sunday morning. As we walked up to the lion enclosure, we were met by a most incredible sight that left us totally humbled. There they were, as if they had been waiting for us, a most magnificent pride of white and golden lions and lionesses.

As we approached, I started to sing to them. They tilted their heads to listen, and I felt honored. We sat with them for a while and then asked them to follow us so we could move to a nice spot in the sun. They slowly got up one by one and followed us along the fence to our spot in the sun.

We started to connect with them and all we got was a bit of small talk with no earth-shattering information. By this time, there was a crowd around us. While Linda and Carrie connected to some of the lions, I sat up against the fence in a relaxed state. I closed my eyes and for a few moments my energy merged with theirs. I was blown away. I felt what it must be like to be such a magnificent beast. I saw myself in their camp with them, walking among them, being a part of them. It was a most serene feeling of royalty, greatness, contentedness and peacefulness. I was privileged to be there with them.

After that experience, we decided to move away for a while, and have some lunch before coming back to them. I thought that if this was the only experience I was going to have on my first meeting with them, then I was happy.

What happened next left me totally astounded.

We came back from our lunch. I had left my notebook and pen in the car and was just standing at the fence enjoying being in the presence of such great beauty. Suddenly, the male lion who was sitting in front of me and looking directly at me, spoke to me.

# TALK TO THE WILD SIDE

"We are not just lions, we have a much bigger purpose. We have a lot to teach, and we have a lot of wisdom and greatness to share. Be open; learn from us, all things will be revealed in time."

I stood there with my mouth open. As I was still connected to the male lion, I asked him for a message for myself.

He said, "You have animal ability, use it in all things."

"Don't get caught up in the rush, enjoy being yourself, relax and just be.

Feel the energy we share with you, be thankful, be grateful, and be happy."

"Be true to yourself, be honest with feelings. You see us as we are; let us see you as you are. Don't pretend to be something else. You are good enough, respect your own importance, we respect ourselves."

Although a short message, it was a very significant message to me, as I have a strong sense of intuition, clairaudience, clairvoyance and clairsentience, which are all the senses animals use to communicate.

I suggested to Linda and Carrie that they see if the lions had a message for them.

Carrie then received a message from one of the lionesses.

"We are here to teach higher learning, we have a higher purpose on a soul level. One of our main purposes is to teach spirituality. Earth is suffering because we have lost our focus, and we need to preserve our earth and our universe, otherwise man will destroy it."

Later Linda shared with us the message she had received from one of the other lions.

"Learn about humility, patience and love. Stillness in your day can create an energy that allows your life lessons to happen with love and not tension. Love every day, and enjoy life."

We left the park after thanking the lions for their messages, feeling totally humbled to have had the privilege of being in the company of such greatness.

Have you ever noticed how energized you feel after spending time in the presence of lions? They have the most amazing healing abilities.

I look forward to my next meeting with them.

### The day we met the elephants

It was a warm winter's day and we sat waiting for the jeep to arrive. We were on our way to meet the elephants at a reserve near Sun City. The jeep arrived and we all piled in.

We drove along a dirt road lined on either side with indigenous thorn trees. As we drove along this winding road, my excitement was building. We went over bumps, around bends, under trees. The dust was all around us. I felt the anticipation building even more as we got closer. At times I could hardly breath. I started to smell the faint smell of elephants in the distance.

Finally, as we rounded the last bend, I looked down into the valley on my left. And there they were, a group of five of the most beautiful elephants I had ever seen. They stood around the dam waiting for us.

# TALK TO THE WILD SIDE

We finally came to a stop, got out of the jeep and slowly started our descent towards them.

We were introduced to the elephants one by one: Sharo, Sharu, Sapi, Chekwenya and Michael. They were magnificent.

I turned to the elephant named Michael and said, "Hello Michael, I'm an animal communicator. May I talk to you?"

He replied, "You must feed me first," at which he lifted his trunk and opened his mouth to reveal a soft pink tongue.

I picked up a handful of cubes, put my hand right inside his mouth and gently stroked his tongue. Michael then went on to tell me that they were all happy and loved their jobs, and enjoyed taking people for rides and chatting to them. However, he told me, Chekwenya, the only lady in the group, desperately wanted to have a baby, and Sapi was bored and needed more stimulation – that's why he was naughty. Sharu was insecure and, to boost his own ego, he was bullying the others.

When I later related these messages to the rangers, they confirmed Chekwenya was indeed trying to get pregnant, and Sapi was a bit lethargic and possibly bored (he was very naughty), and that Sharu was indeed a bully.

I sat in the presence of these great beasts for a while after everyone else left. I just enjoyed being with them.

I've always had a special affinity to elephants, as one of my spirit guides is an elephant. I will be going back to visit the elephants and perhaps will see a baby for Chekwenya…

## 🐾 The young zebra

I had an interesting experience recently with a zebra.

# TALK TO THE WILD SIDE

I was going to a talk at the Voortrekker Monument in Pretoria, and as I was driving up the driveway I noticed some buck and zebra at the side of the road. I decided to stop and chat to them. I got out of my car and walked over to them. As I got close, a young zebra stood out from the group.

I said to her, "Can I talk to you?" To which she looked me right in the eye and replied, "No! I don't know you."

She then turned around and walked off.

I realized much later that I had forgotten to introduce myself to her, which is something I usually do when connecting with a strange animal. It is the same as walking up to a stranger in the street and starting a conversation – some people will talk and others not. Animals are the same.

## 🐾 Venus

I had been taking part in a course at the Lanseria Lion Park connecting with nature. As part of this course we would connect to the lions to see what they would teach us about our own lives. I was asked by Gail, who was running the course, to spend some quiet time in the cub enclosure. I was delighted but had no idea what treat lay in store for me.

As I entered the enclosure I noticed four cubs sleeping in the sun. I looked closer and saw that these were the same cubs I had worked with a couple of months earlier, when they were only one month old. They were now strapping four-month old cubs.

Venus was the one I had spent two days connecting with, while facilitating on an Eco Access course with some blind children. I was very excited to see Venus again and I decided to work with her. I sat next to Venus as she slept. Every time someone walked past her and stroked her, she would hiss and snap at them. When the other people eventually left, I said to Venus, "Venus, I'm going to stroke you now, is that okay?"

I stroked her and she was totally relaxed, with no hissing or snapping.

After spending some time with this beautiful little cub, I lay on the ground about a meter in front of her, on my elbow, and said to her, "Don't you want to come and put your paw in my paw?"

I waited.

After about five minutes she got up, walked over to me, tapped my hand with her paw, climbed up onto me and put her paw around my shoulder and her cheek on my cheek. She then slowly climbed off me and walked over to the other cubs.

A lion cub had just hugged me. I was blown away.

##  Diamond

Later that same day, we went to connect with the big lions. As we got to them, all the white lions, as always, came up to the fence. There was one golden lioness that sat back from the group. She was relatively new, still very nervous and not used to people. She would hiss and launch herself at us. I went and just sat near the fence connecting with her. Eventually she settled down a bit, but would still snarl if I moved too fast.

That evening, in a meditation, I asked her for a symbol for me to give to her. I saw a diamond, which settled in her heart centre and her forehead (third eye). I decided to name her Diamond.

The next day I went to visit Diamond. When I got there, she separated herself from the pride and walked over to me. She recognized me. All the other lions were in the corner chatting to the other people on the course, but Diamond sat at the gate next to me, quite relaxed, while I communicated with her. What a gift the lions had given me.

On another of my visits to Diamond, I got there to find her and a lioness I had called Special Star, lying apart from the rest of the pride. I sat down, looked at Diamond and said, "Diamond, won't you come to the fence so I can sit next to you?"

After a while she got up and walked up to the fence to see what the others were doing. She came straight back to me, rolled on her back in front of me, and relaxed so that I could spend time with her. Her energy was so strong I could feel it right through my entire body. The love that I felt pouring out of her made my whole body tingle, and at that moment nothing else mattered, I was with my lion heart lady Diamond.

When communicating with any animal, it is important to feel all the sensations going through your body. This is part of the way they communicate with you. They show you through your own body where the pain is, or transfer their love to you by giving you a feeling of total love. You need to become aware of changes in your body temperature or moods, sensations or feelings. The more in tune you are with your own feelings, the easier it will be to feel the moods and feelings of the animals.

### 🐾 Receiving energy from Jade the jaguar

Jade is a very elegant lady jaguar who we have been visiting at the Lion Park. The first day that Linda, Carrie, Gail, who was running the lion courses, and I went to visit Jade, we sat at the fence to watch her.

Jade saw us and quickly came over to us. As we watched, she

paced up and down, just in front of us. I became aware of her energy, her aura, which she was spinning off herself and onto us. As Jade continued to pace, we were getting more and more of her energy. When she thought we had had enough, she slowly walked off and left the four of us very energized.

When you work with wild life, you become aware of how they mirror your own feelings. If a lion is restless and paces in front of you, ask yourself what it is trying to tell you. Are you restless, are you pacing over decisions you need to make? What is going on in your life that the lion is trying to make you aware of? If the lion is aggressive, then ask yourself what you are scared of. No animal is naturally aggressive. Aggression comes through fear or pain. So, if you are faced with an aggressive lion, ask yourself whether you are scared of the unknown, what fears you are hiding.

# TALK TO THE WILD SIDE

I would like to share with you a poem that was written by a very close friend of mine, Keli Smith.

## The Dream

Fast asleep one night
In the deep twilight
A lion came to me
He said hello and off did go
I awoke and did not see

So he came again
This time not tame
Exploding through the grass
Standing alone I did know
This breath will be my last
As I stood accepting death
My body calmed to quiet breath
He stopped in front of me
And said "do come and sit in the sun"
We will talk and you will see

And many things to me he said
Until the sun was low and red
And I sat in his golden light
When he had to leave, I did grieve
And remembered my dream that night

So I connected to the Lion Heart
And they truly became a part
Of my very being
And I understood the thread that lead
They were calling me

As my heart beats in my breast
Never again will my light rest
Their message is plain to see
To find my light on the darkest night
I must understand the Lion is me.

# TALK TO THE WILD SIDE

I decided to go and sit with Sly, a huge male lion, and meditate with him.

When I arrived in his camp, he was waiting for me at the gate. I drove in and Sly turned around and started walking in front of me, leading me in. We went down a dirt road and around a bend, as he took me to a shady spot under a tree. As I watched, he circled around and around in the middle of the road, as if preparing the ground and spreading the energy. Eventually Sly lay down. I sat quietly and started to meditate. I meditated with Sly for about 25 minutes and, when I eventually opened my eyes, he had moved off to join the rest of the pride.

He told me that he and all the other lions would walk side by side with me down my spiritual path, preparing the energy and keeping me strong and focused.

In my meditation I had seen a magnificent gold sun with gold rays sparkling all around, and at the tip of each ray was a lion. I was in the centre of the sun and all the lions were forming a gold circle of protection around me.

Their message was that in everything you do, do it through your heart. This is how the lions work, they don't stop to think, plan and analyze, they react from their heart.

There is great respect and a very special bond between the predator and the prey. When a lion goes after its prey, it will ask its prey for permission. While the prey's physical body will react out of fear, the soul will feel a great honor at sacrificing itself. At the moment of death, the two souls will merge, and for a moment the predator and prey will become one.

Animals don't think of death the way we do. They are far more accepting and understanding of it. They feel that if they are killed out of respect, for food or the survival of others, this is easier to accept. But if they are killed for sport or profit, as in canned hunting or trophy collecting, this distresses them.

The Bushmen would always ask for permission from the animal they were hunting, and explain the reason why they were doing this. They would always say a prayer and honor the animal before hunting it. It was done out of total respect.

### The cheetah

I was with a group of people driving through the Lion Park when we stopped to look at a magnificent cheetah sitting up high on top of the roof of her shelter. I hadn't gone planning to communicate, but felt a strange pull and got out my pen and paper. As we started driving away, the message came fast and I started writing. This is what she said.

"The beauty you see in us, is the beauty we see in you. Many of you have lost your way, we can help you find your way back to your true self. Be flexible; don't judge, for we are your eyes. Trust in yourself, be beautiful."

And then we were gone and I was left reflecting on the magnitude of this message.

### Ian's story with the lioness and her four cubs

Ian Melass, the manager of the Lanseria Lion Park, was doing his rounds checking up on all the lions he loved so much and was so close to. As he came to a camp right at the end near a wooded area, he stopped in amazement. There was Bijan with four tiny little cubs. No one had realized she was expecting, she hadn't shown any signs.

# TALK TO THE WILD SIDE

A lioness will never let anyone close to her while she is feeding her cubs. She will show great signs of aggression: spitting, hissing and snarling, even attacking. However, Bijan was totally relaxed. She allowed Ian to look at her babies. When he had finished, off he went to spread the news.

A couple of days later, when Ian was checking up on the cubs, he got to their enclosure and was dismayed to find no lioness or cubs. He slowly walked around the outside of Bijan's enclosure in a wooded area and turned around with a jolt. There she was lying in the bushes outside the enclosure with her cubs. She had dug a hole under the fence and was lying very relaxed in the bushes. At that moment, she got up and slowly started to walk over to Ian who thought it was now his last day on earth, but stood dead still, just as he has always warned us to do if approached by a lion.

Bijan walked up to Ian and lay down. By now he had summoned some help and, with the aid of some food, slowly coaxed Bijan back to her camp and safety. But the cubs were still on the outside. So Ian, being the amazing man he is, picked them up, one at a time, and handed them to Bijan, who gently took them out of Ian's hand, one at a time, and carefully placed them on the ground in front of her. All was well in the lion camp.

A few days later, I went on one of my regular visits to the Lion Park. When I saw Ian, he said, "Come and see, I've got something to show you."

He took me to see all the lions. We went for a walk along the front of the camps, around the corner and into the wooded area. There in front of us was a magnificent lioness with four tiny little cubs.

We walked up on the other side of the fence and sat down on the ground, right in front of Bijan and her four cubs. We were less than a meter away from a lioness feeding her cubs and she was totally relaxed.

# TALK TO THE WILD SIDE

As Ian explained to me, I would never again get to experience being so close to a lioness and her cubs.

This lioness is a very special lady and her cubs are amazing.

Thank you, Bijan, for allowing me to share this incredible experience with you.

## 🐾 The Johannesburg Zoo

On a Tuesday in January 2005, I went to visit the animals at the Johannesburg Zoo. It was lovely. I was the only person there except for the odd child or adult. It was so peaceful.

While I was walking down these shady lanes with different animals on either side, I came across the elephant camp. As I rounded the corner, there were two beautiful elephants right near the fence, and there was no one else around. I walked up to them and sat down, to just spend time with them. One of them was a male, big, beautiful and sturdy. He said he was called Henry. The other one was a very pretty lady, whom Henry told me to call Sadie. They were both very proud elephants.

While we were sitting chatting, a family walked past with two babies in prams. Henry and Sadie immediately took note and watched with great interest as they went by. Many elephants had told me in the past of their love for babies. So I asked, what was it about babies that they loved so much?

"It's their openness, they are unconditional beings, they accept without question or judgment. They give off great amounts of love. They are uncontaminated by life at this stage and we love them as they love us."

# TALK TO THE WILD SIDE

### 🐾 Alan and his rhino ordeal

My very close friend, Keli, and I decided to take our respective husbands, Les and Alan, on a visit to The Lion and Rhino Park near Lanseria.

After driving through the whole reserve, we decided to go into the nursery to visit the baby animals. There were lion cubs, tiger cubs, various birds, including cranes, a baby zebra and a rhino.

We were walking down a path and, while I had turned my head away for a split second, I heard a noise behind me. I quickly turned around. There in front of me was Alan, my husband, parallel to the ground, about a meter off the ground, with a rhino between his legs. I quickly said, "Get off, don't sit on her head, she's a baby." She was a big baby, with a big horn. Alan was a bit bruised but the rhino, whose name was Clover, was fine.

The moral of the story is, "Start acknowledging animals and communicating with them." They don't want to feel ignored or taken for granted. They want to be appreciated for who and what they are. If you don't acknowledge and appreciate them, then some day you, too, could find yourself with a rhino between your legs. Take care.

### 🐾 Our Lion Park visit in March 2005

My mother and I took my friend, Wynn (Mishka's mom), and her grandson, Michael, to visit the lions. Michael was out from England on holiday at the time.

We first took a drive through the park and saw the zebra, ostriches, some buck, and of course Purdy and Gambit, the giraffes. We drove past the spotted hyena and the cheetah. We then made our way to the lion enclosures and, on the way, my mother said to me, "Are these not the lions that eat the tyres?"

# TALK TO THE WILD SIDE

I replied "No, it is the hyenas that eat the tyres."

As we drove into the first enclosure, there was Letatsi, a huge male white lion, and a lioness he was busy mating with. We drove through the gate and immediately the lioness started to chase our car. We drove slowly to the end of the road.

My mother, slightly panicky, said, "What if she breaks the glass and gets into the car?"

We then noticed a big ditch in the road ahead. There was no way out – we would have to back up. But alas, there was a lioness behind our car and by now Letatsi had come over to see what all the fuss was about. We had a big lion in front and a big lioness behind.

Nowhere to go!

Suddenly, there was a big thump and my mom asked, "What was that?"

I said, "We now have a lioness on our roof."

"What do we do now?"

To which I replied, "Nothing."

With that, the lioness jumped off, all four paws on the ground. Everyone in the car breathed a sigh of relief.

But all was not over. By now Letatsi wanted to get in on the act. He walked slowly past every window of the car and stared at us as he went. I turned around to see where he was going and there he was, his face filling the entire back window and he stared at us with his pale blue eyes. He gave the car a few swipes with his giant paw, and then we heard a crunching sound.

Again my mom said, "What is he doing?"

"Eating the car," I replied.

"Tell him to stop, tell him to stop," she said.

With that, off they went, and we could get back on the road and make our way safely into the next camp.

### Visiting the cheetahs

In April of 2005, my partner, Sandy, and I were invited to go and visit some cheetahs at a cheetah sanctuary near by. We were met by one of the ladies that work there. She explained to us that one of the female cheetahs was very aggressive and kept biting people. With her being so unpredictable, there was only one person who would go in with her.

I went and sat next to the fence of her enclosure. She came up and lay right against the fence where I was sitting, and I started to communicate with her. I found she was very insecure and jealous. I explained to her what was expected of her as far as behavior was concerned, and she seemed to relax.

Two weeks later, I received a phone call to say that ever since that day she had been much more relaxed and gentle. She had calmed right down, and now would even take food gently out of her handler's hand without snatching.

### U Shaka

After a long trip round in circles, my partner, Sandy, and I finally found the aquarium in Durban. For quite a few months I had been feeling a pull towards the aquarium and the sharks. I knew there was one shark in particular that wanted to connect with me.

We walked all over and saw many different types of fish. There were lots of sharks. I walked around looking and waiting for the shark to come forward. As we were about to leave, a shark started to approach us. I said to Sandy, "This is the one."

I know nothing about sharks but I felt a very strong female energy, and thought this shark must be a female. She swam right up to the glass next to me and I connected to her. The message she gave me was short but very profound.

"People fear what they don't understand, they see us as ugly in their minds and, because of this, they fear us. We are not here to kill people, we kill out of necessity and hunger and self-protection. It is our instinct. We are really very gentle and misunderstood."

The whole time I was chatting to this beautiful shark I felt a gentleness and a sadness. The message she was desperate to get across to us was to look deep within ourselves and not to judge appearances. Be kind to ourselves and study more about the things we don't understand, or we fear. We need to love ourselves and all things.

## Connecting to whales

I had been running quite a few workshops and doing lots of private consultations and photograph readings. I was exhausted. I decided to go to Keli for a healing to restore some of my energy.

Just before I got there, Keli had been walking around her garden thinking about what music she should play during my healing session. Suddenly she got a message. Play the whale sounds.

I arrived shortly afterwards, got up onto the healing bed, and started to relax. Keli put the whale sounds on and started with the healing session. As I was relaxing and listening to the whales cavorting in the ocean I suddenly felt myself connecting to the whales. I became aware that there were three whales. One was an elderly whale and the other two were young and in love.

I just lay there enjoying this contact, not trying to communicate. Then the whales suddenly spoke. They said that they were from another dimension and were here to help humanity. They were our teachers. They wanted us to know that if they had finished their work here, or were wanted at home in their dimension, they would be summoned back. One way they could return to their home was to beach themselves.

Many people think the beaching of whales is unexplained, but according to the whales it is their way of returning home.

I found out later that two weeks earlier some whales had beached themselves in Cape Town and people had been frantically trying to help them.

### Bird on a broom

Early yesterday morning I was working in my office when I heard a loud banging noise. I got up to go and investigate. There was a little finch, it had flown in and couldn't find its way out.

It flew around the entrance hall, into the kitchen and finally settled on the face brick wall in the dining room. This little bird was obviously in distress, its beak was open and it was panting heavily. I stood for a while just sending it some calming energy, giving him some time to calm down before communicating with him.

When he had calmed down a bit, I said to him, "Hello little bird, I'm your friend and I want to help you get out where you will be safe. You need to trust me so I can help you. I'm going to fetch a broom, please wait calmly for me."

When I got back with the broom, he was still sitting on the wall waiting for me. I raised the broom up to where he was and said, "This is the broom, I want you to climb onto it so that I can take you outside where you will be safe."

After a couple of seconds, he hopped onto the broom and I slowly lowered it, and gently carried him out to a tree, where I placed him on a branch.

I left him for a few minutes then went to check up on him. He was still there on the branch.

"This is my finger, I'm going to scratch your chest," and I slowly reached up and scratched his chest.

"You are now safe, you can fly away to join your friends."

With that he flew off to join his friends that had been watching with great interest.

All it needed was for me to be totally calm and relaxed, and to explain the steps I would take to assure the little bird's safety. If I had run around panicking, he would have panicked and I would not have been able to communicate with him successfully.

### 🐾 Talking to a cobra

One Monday after taking Riff-Raff and Stacey to dog school, Thomas and I brought them home and let them out of the car. They ran straight through the carport and into the kitchen. I got them settled, called Kate (Thomas's wife) and her little three-year-old daughter, Jenny (named after me), to come so that we could go to the shop. While Kate was putting Jenny into the back of the car, I heard a bloodcurdling scream. Kate and Jenny ran back into the house.

I got out of the car to see what the fuss was all about and there, in the corner of the carport, next to the kitchen door, I saw a cobra. My first thought was not "Eeeeeek! Snake!" but rather, "Wow, what information can I get from this little snake for my book?"

I asked Thomas to go and fetch me a box. While I was waiting for him, I looked at this beautiful little snake. She had a silver glow about her and I could feel her fear. She was young.

Thomas arrived, and I put the box on the ground in front of the little snake. She was very tense. She was up, her hood was out and she was ready to strike.

I said to her, "You are safe, I'm not going to hurt you, you need to climb into this box so that we can put you out in the veldt where you will be safe."

After a few minutes, I could feel her starting to relax. She flattened her hood and slowly slithered into the box. Thomas closed the lid, picked it up and took it into the veldt where we released her, and she slithered off to safety.

###  Talking to flies

A friend of mine took part in one of my very first workshops. This workshop was held over two weekends. After the first weekend, she went to a gymkhana. As everybody knows, where there are a lot of horses, there are usually flies. During lunch, the flies were everywhere and irritating everyone, so she decided to use a technique she had learned on the workshop. She pictured a giant fly swatter above her plate of food. When she looked again, her plate of food was the only plate with not a single fly. All the other plates were covered in flies.

When communicating with any animal or insect, if you ask them to go you need to give them somewhere to go. For instance, if you have a problem with termites, put a fallen log, or piece of rotting wood in the bottom of your garden. Ask them to please leave your house and move into the log or wood that you have given them at the bottom of your garden. It doesn't help if you just say, "Leave my house." They need to have an alternative.

### Hungry snails

A friend of mine told me she was having a lot of problems with the snails eating her cabbages in her cabbage patch. I suggested she give them a cabbage of their own. Also to tell them that if they get caught eating from the other cabbages, she would have to kill them – but they could eat off the cabbage she had provided for them.

When she came out the following day, all the snails were busy eating the cabbage she had given them. They were leaving the other cabbages alone.

I told her that when this cabbage was finished she should give them another one. She hasn't had any more problems with the snails. Everyone is happy and living in harmony.

# You are My Light

We live our lives together
In good times and in bad
You're my light, you're my friend
The best I've ever had

You always are so trusting
so happy when you see me
You're my light, you're my friend
There's only us, there's only we

In dark I reach to touch you
Your presence always here
You're my light, you're my friend
My darling pet I hold so dear

You fill my life with love and joy
You mean the world to me
You're my light, you're my friend
May I then too, your best friend be

My love for you just has no end
I learn so much from you
You're my light, you're my friend
You're my teacher too

And when it's time for parting
and pain and heartache wait
Let's be the light, let's be the friend
That's waiting there at heaven's gate

*Alba Delport*

# CHAPTER Eight

## THE HEALING EXPERIENCE

*Be kind to your fellow man
and us and we will help you
soar to great heights,
it is our promise to you.*

**Red**

# THE HEALING EXPERIENCE

Many people ask, "Why do animals get sick?"

There are many reasons for this. The same is true of their human companions. If the aura, which is the electromagnetic field that surrounds the body, is weakened or torn, this could lead to disease manifesting in the physical body. There are many reasons why the aura gets damaged. Some are in our control, while others are out of our control.

Things that contribute to a damaged aura, to list just a few, include:

- prolonged use of drugs or medication for ailments like diabetes, arthritis or heart disease, as well as various forms of pain medication
- long-term exposure to physical abuse
- exposure to negative thoughts and energies, and
- second-hand smoke.

A lack of harmony in our pets' lives produces a continuously stressful existence for them. Trauma to any part of the body can derail the life-force energy within the animal. This not only affects the physical body on a cellular level, but also affects the etheric body on a psychic level.

Very often you will see that a person who is being treated for arthritis will have a dog with the same condition. This is because animals have the ability to take on their human companion's problems. The problem is very often seen in the animal before being identified in the human.

Since the aura supports the physical body, it is necessary to keep the aura strong and healthy. The etheric body is the layer of the aura that is closest to the physical body. By picking up and treating a disease in the etheric body, you will prevent the disease from

# THE HEALING EXPERIENCE

manifesting in the physical body.

You can train your eye to see this yourself. Ask a friend to stand in front of a blank wall, focus on the area around their head, relax and concentrate on your breathing. Within a few minutes you will start to see a silver shadow form around their head. This is the etheric aura. With lots of practice, you will be able to start seeing colors, which are the other layers of the aura. A good book to help you develop this skill is "Auras: See Them In Only 60 Seconds", by Mark Smith.

A very important part of a healing session is the balancing of the chakras, which are the energy centers within the body. There are seven main chakras and each one relates to a different area or organ in the body.

- The first chakra is at the base of the spine and is seen as red
- The second is in the sacral area and is orange
- The third is in the solar plexus and is yellow
- The fourth, the heart center, is green
- The fifth chakra is in the throat center and is blue,
- The sixth is the center between the eyes and is indigo, and
- The seventh chakra, at the crown of the head, is violet.

If any of these energy centers are not balanced, the whole body will be out of balance, which could result in disease.

Animals don't have the ability to process stress the way we do. For this reason, learning to telepathically communicate with your pet can and does have great benefits. Most disease stems from deep emotional trauma. By telepathically communicating with your pet, you can get to the root of the emotional problem and be able to understand it and deal with it more effectively.

# THE HEALING EXPERIENCE

## 🐾 Victoria

Ashley brought Victoria to see me. Victoria was a bulldog who had been bitten by a tick and developed a very serious bout of cerebal biliary that had severely damaged her brain. She couldn't recognize anyone, struggled to eat and had difficulty walking. She had been to a number of vet specialists who had suggested she be euthanased.

To Ashley this was devastating. Victoria was her baby. She phoned me and asked if there was anything I could do to help. I suggested bringing her over to see what we could do.

Before Victoria got to me, I started to communicate with her. By the time I met her, she already seemed to recognize me.

I started to do Reiki healing sessions on her, and found most of the damage was to the right side of her brain. I carried on working on her for the next three weeks, at which time Victoria once again became a normal healthy dog, who would live a happy life for the next four years.

## 🐾 Healing through communication

I went to see a dog recently that was having a lot of trouble with an upset tummy and had become bad tempered. In other words, she was just not happy. I had a chat to her and simply listened to all her anxieties. She was very insecure and was carrying a lot of guilt. She told me that the man of the house was going through a lot of stress at work, and was smoking a lot. There were a lot of arguments and temper tantrums in the house. She felt it was her fault for causing the stress and had become guilty. All I did was listen to her and reassure her of her importance in the family, and the fact that it wasn't her fault.

# THE HEALING EXPERIENCE

Overnight her tummy trouble cleared up, she stopped being bad tempered and was her old happy self again.

The wife later confirmed her husband had started smoking a lot, was suffering from stress at work and would come home in a bad mood at the end of the day.

## Some common issues

- A dog lifting a leg on the furniture is usually because it needs attention or because a new addition to the household is causing insecurity or stress.
- Aggression in any animal is usually due to either fear or pain.
- A dog digging in the garden is often an indication of boredom.
- A cat scratching the furniture is simply because cats need to scratch. If you watch a cat scratching, it digs its claws in to the object and stretches its whole body. In doing so, it manipulates its entire spine. So always provide a suitable scratching post for your cat and explain to your cat what the scratching post is there for.

Some people have their cats de-clawed to protect their furniture. This is an extremely cruel and inhumane act. The cat then has no way of defending itself against danger. It also has no way of manipulating its spine. Most damaging of all, is that this operation is one of the most painful a cat can go through, since the whole joint is removed – the equivalent of having your fingers amputated. The memory and trauma of this can last a lifetime.

Telepathic communication can have dramatic results for pets suffering from separation anxiety. By identifying the causes, solutions can be found to enable healing on the emotional level.

# THE HEALING EXPERIENCE

If more people learn to do this form of healing, and more animals can be helped, isn't it worth it? We receive so much unconditional love, isn't it time we returned the favor?

### 🐾 Kerry with hepatitis

I was busy running a healing course for animals and we had all gathered round. In the group were my partner, Sandy, nine other ladies, and myself. We were just about ready to begin when the phone rang. It was a lady asking us to help with her little Maltese Poodle, Kerry, who had severe hepatitis and was dying. She was rushing home from work and didn't know if Kerry would still be alive when she got back.

I gathered everyone around and we sent group distant healing to Kerry.

The next day, Sandy received a call from Kerry's owner to say that as she had got home and opened the door, Kerry ran up to meet her. She couldn't believe it. Kerry was fine and, months later, is still doing well.

# CHAPTER Nine

## HORSE SENSE -
## the horses have something to say

*When you look at us,
open your eyes and see us,
appreciate our finer points -
we are there for you.
Learn from us and we will learn from you.
We must work as a team to better the world.*

**Penny**

# HORSE SENSE

### 🐾 Horses at the fence

Snoopy had cornered me for a chat, so as a result I was running late for an appointment. I was driving down a road in Apple Orchards, the road was lined with trees on either side. As I came around the bend, I heard voices in my head.

"The communicator, the communicator, the communicator."

I thought to myself, "Wouldn't it be funny if all the horses were lined up against the fence looking into the road waiting for me"

As I got to the end of road, there they were, three beautiful horses standing at the fence looking into the road.

I realized never to underestimate the abilities of animals.

### 🐾 Rebecca

Rebecca is a big beautiful, almost black thoroughbred mare that lives with me. The more I work with her, the more I realize she is a very wise lady and an old soul, which means she has had many lives here on earth. Rebecca has so much to teach us. I sat with her one day to see what she had to say. This is her story.

I asked Rebecca, "Do you have any information for me to include in my book that will help people understand things from a horse's perspective?"

This is what she said.

"For many centuries and lives, horses have been thought of as beasts of burden. Although we do carry the burdens of our people, we are very evolved beings. Although we can learn a lot from humans, we can also teach them a lot if they would just listen. I'm saying this not out of judgment, but out of concern. Humans have shut off their abilities by not listening to or feeling our thoughts. Humans need to still their minds and focus on their lives and worry less about what others are doing.

"We horses are very sensitive and many people treat us not as beings with knowledge or feelings, but as machines with which to relieve their own stresses. We accept this, knowing that one day humans will wake up and see our sensitivity and understand our concern for them. I feel honored to be included in this book, I will at last get to spread my knowledge to others."

I asked Rebecca, "What do you feel about all the other horses in your herd?"

She said, "If you notice, with all the work we have done while living here with you, teaching communication skills and healing, we have greatly improved the lives of many people. We have raised their level of awareness and consciousness and, in turn, our own. We feel privileged to be able to teach in this way. We are doing what we were sent here to do."

I thanked Rebecca for this amazing message and she walked back to join the others in the paddock.

### 🐾 Red

Red walked over to me and I heard the words, "You must talk to me, I've got something to say."

Red is one of Rebecca's companions. He is a beautiful, but stubborn, chestnut horse that came to live with us after we lost Rebecca's best friend, Topaz. She needed another horse she could bond with, so Red came to live with us.

I asked, "Red, what do you think of what Rebecca told me?"

He said, "It's all true, its all true, Rebecca is a very wise lady and an old soul, she has taught all of us much, and her wisdom is never ending."

I asked him, "Red what have you got to add to my book?"

"Just want to say when I came to live with you I was a baby, inexperienced in matters of communication with humans. I had never experienced healing. I have now found myself and been given the opportunity to grow and develop.

"There are many horses out there that, due to stress and abuse, have forgotten what their purpose is. All it will take is someone to point out the important role that horses play in the relationship between horses and humans.

"We can help bring out your own power and freedom so that you, too, can develop. There are many changes coming. For far too long people have been asleep. It is now time to wake up and take charge of your own lives. That's all I have to say, beautiful people."

### Penny's book

The day finally arrived. It was hot and steamy with the slight sound of thunder in the distance. Two months previously I had asked Penny, one of my miniature horse companions, if she had anything to add to my book. I had told her that I would come and sit with her to get her input as soon as I had time. I then got very busy with all my workshops, talks and various articles I was writing for magazines. In the meantime, Penny was going around telling everyone that she was writing a book.

# HORSE SENSE

Finally the day was here, I walked out into the paddock where Penny was waiting for me. I brought her in to the stable yard where we wouldn't be disturbed. I sat under a tree and started to chat to Penny.

She said, "At last, I thought I would have to get someone else to help me write my book."

I said, "Sorry it took so long, but it is actually my book."

She said, "Enough of the small talk, let's get on with it."

So we started. I asked Penny, "What would you like to add to this book?" No reply, I tried again. "Penny, what is the one thing you can tell me that will help us communicate more effectively with animals?"

Then the floodgates opened and this is what she said.

"Relax, observe, listen, you hear something, you feel something, you think something and then you ignore it. You say, 'It's only in my head', well how do you think it got there? We put it there. Listen to your feelings. You call it a gut feeling; we call it a head feeling. We transfer our feelings first into your head and later into your gut. Listen with your whole body.

"Learn about yourself. Hear what your body is telling you. Be in tune with your own feelings so that you can differentiate between our feelings. The key to happiness and good communication is love. Without love nothing exists.

"You have taken on a great task. Before you can learn to communicate with animals, you need to learn to communicate with your fellow humans, with love and without judgment. Be truthful and appreciate all you have and don't dwell on what you don't have. You need to love yourself."

# HORSE SENSE

I thanked Penny for her amazing messages, gave her a treat and watched as this very wise little horse went off to join the others in the paddock.

I was thankful to have such amazing animals sharing my life with me.

## 🐾 Ballinor

I asked Ballinor, one of my miniature horse companions, and Penny's friend, "What are your feelings about animals in captivity?"

He said, "It's a state of mind. What you call captivity is only a part of it. It's true animals do need a certain amount of space and sunlight for their physical and mental development. They also need stimulation, to relieve their boredom (mental and physical). When people look at an animal in a cage, they see the cage and feel how they would feel in a cage. Animals don't have human thoughts. People tend to humanize the animals.

"Remember humans are humans and animals are animals. They don't judge or criticize, they accept situations more readily. What does concern the animals is abuse. We don't understand why, when we are so full of love for our human companions, some of them treat us so badly, with no love or respect, and think of us as inferior.

"Communication is a two way street, it's easy to talk but the secret lies in being able to listen. Have you ever tried to have a conversation with someone who doesn't listen? It's frustrating."

I asked Ballinor, "Is there anything you don't like that upsets you?"

He answered, "I don't like the anger people have for each other, and they need to learn to get along with each other for the sake of all humanity and the universe."

## 🐾 Tootsie

Tootsie is another one of the miniature horses that live with me and she has the sweetest nature.

I asked her one day, "Is there anything in particular that you want to talk about?"

"Ummm, let me think.

"Yes—the sensitivity of animals.

"What some people don't realize is that their stresses, moods and even illnesses affect the animals in their lives. Animals work as emotional buffers. Like sponges, they absorb all the negative energy that bounces off the arguments their people might be having in front of them. If our people are sad we feel it, if they are angry we feel it, if they are depressed we feel it and it effects us. So, for this reason, my plea to you is: look after yourselves physically and emotionally for the sake of the animals in your lives that love you so much. The healthier and happier you are, the healthier and happier your animal friends will be."

I then asked Tootsie, "What is your biggest love or fear?"

She said, "My biggest love is the love I share with the humans in my life and the children that come to visit.

"My fear is that people won't wake up and start to communicate with all things. That they will stay closed and hamper their own spiritual growth."

# HORSE SENSE

### 🐾 Tales from a donkey

We got out of the car at The Highveld Horse Care Unit in Meyerton and looked around at the beautiful countryside. The paddocks were immaculate. The horses, despite the fact they had all come from some form of neglect or abuse, were happily grazing. Then we saw Lucy, a beautiful young donkey, about one-year old, standing in a paddock by herself.

A few months earlier, I had started running an ESP animal healing course, the first of its kind in the country. There were ten people on this course, and it was being run over a period of four months. This was their final day, and I had brought them here to do an exam. They were going to communicate with the horses, as well as do a healing session on them.

We walked over to Lucy who was very timid at first, and didn't trust anyone. She had been at The Horse Care Unit for five days, and this is her story.

"I grew up in a township. Every morning, I would get strapped into a cart alongside my mother. We would pull the cart all day until evening. It was a hard job but necessary, it was our job.
"One day while we were doing our job, a motorcar came around the bend at full speed and crashed into us. There was a lot of noise, I was afraid. My mother lay there not moving. Then I heard someone say she was dead. I couldn't believe it. She was gone. What was I going to do?

"Some nice people arrived and took me away. I wanted my mom. Where was I going?"

That was when The Highveld Horse Care Unit rescued Lucy and took her to a place of safety.

# HORSE SENSE

I arrived five days later with my team of animal healers and decided to adopt Lucy for The Animal Healing Centre.

Two weeks later a horsebox arrived at my plot and the door opened and out walked Lucy. She was a bit unsure of what was expected of her, and why she was here. I walked her over and showed her her new stable, complete with teddy bear. I introduced her to the other horses and left her to find her feet.

Later, I explained to Lucy that she now had a new career teaching people to communicate with donkeys. She was very happy. At last Lucy had found her purpose in life.

Then the day finally arrived; it was Lucy's first day at work in her new career. She would now be helping me on my workshops, teaching people to communicate with animals, and letting them know what the world was like from a donkey's point of view.

I had already given a talk on animal telepathic communication. We had also done a meditation and some exercises, chatted to the dogs and cats, and now it was time to go and see what Lucy and the horses had to say.

Lucy had spent the past week very excited about all the people who were going to come and talk to her. By the time we got to her, she could not control her excitement anymore. She trotted up to everyone and greeted them one at a time. She nudged their backs and nibbled them on the back of their legs. "Hurry up!" she said, "Talk to me."

I had asked Marika Schumyn if she would talk to Lucy as she had done some of my workshops before. Marika had a lot of experience, and Lucy was new at this game of communicating with humans.

She started off with some basic questions and found out Lucy's favorite activity was to be free to run around the paddock with the other horses; she didn't want to be hitched to anything like a cart; and her best friend was Red. Lucy said she had finally found her soul purpose, which was to teach and be a friend to people.

Marika asked Lucy, "Lucy, am I on the right path?"

Lucy replied, "Yes, you are a healer, use it. Go forward with confidence."

Lucy said her job was to teach people that if something doesn't work out, they shouldn't give up hope.

"Pick yourself up, dust yourself off, and try again. Things work out in the end. Just listen to my life story and learn from it. Even if you lose some one dear to you, they are not gone forever. With love, you will still be able to connect to them. The secret is not to forget that."

"Be aware of all your own thoughts first. Then it will be easy to differentiate between your thoughts and that of the animal you are communicating with."

As I write this, I see her happily grazing in the paddock with the rest of the herd. I'm proud to have her as part of my team

### A healing story

I had been doing professional readings for people for a few months when I got a phone call. The woman on the other end of the line was very distressed. She asked if I could help her – her horse had suddenly become vicious and dangerous. He had reared up in the paddock and broken her nose. Now, every time anyone went near

# HORSE SENSE

him he would rear up at them. She loved her horse and she needed to know why he was rearing up, and what could she do for him.

I made an appointment to go and see the horse, and a few days later found myself driving the long distance to Kyalami to chat to him. For the purpose of anonymity we will call him Fred.

I arrived at the stable yard an hour later. Fred was standing in a paddock with his stable friend, a beautiful dark bay mare. He was stunning, a deep chestnut color, standing about 16 hands high. I introduced myself to a young couple who said they were Fred's owners. They asked me if I would like them to put Fred in his stable so that I could talk to him. I told them that if he was happier and more relaxed in the paddock, then to leave him in the paddock, but to possibly move his friend to the next-door paddock so we wouldn't be disturbed. This they did, and we were ready to start.

I stood for a while with Fred, just getting him to relax with me before we started. After ten minutes, Fred was ready to tell me his story.

He told me he had come from Zimbabwe. A lady had rescued him, but she had later abandoned him again. He felt insecure and loved his home, but he didn't want to be abandoned again, and was scared. He told me of all the fighting, the noise, the chaos and the stress, which he was glad to be away from, but where was the lady who had rescued him all those months ago?

He said the lady he lived with now was scared of him and he didn't understand why. What was wrong with him? Didn't they love him? I reassured him that they loved him very much, that was why I was there, to help them to help him, and to understand him better.

I asked Fred to tell me something I could tell his owners that would confirm to them that I was indeed getting messages from him. He told me he had some problems that they already knew about, but there were other problems that he wanted them to be aware of. He had sore joints in his front right leg and got very stiff sometimes. His teeth also ached. And he said that he got bored in the school but loved going on outrides – but only if his friend went with them, since he didn't want to go alone. However, his owner was scared and didn't want to take him out.

When we had finished our chat, I thanked Fred, told him how much he was loved and how special he was to everyone. Then I went off to relay the messages I had received from Fred.

The owners were amazed. A lady who had put him in a big paddock and fed him every day had indeed rescued Fred from Zimbabwe. Alas, she didn't spend any time with him and this made him feel abandoned. They told me they were afraid of him and didn't take him for outrides for fear of what he might do. For this reason, they worked him only in the school.

They confirmed that he sometimes got a very stiff front right leg, but didn't know anything about his teeth being sore. They later got the vet to look at his teeth and found an old abscess, which was treated. I gave them a few suggestions on how to improve Fred's confidence and told them how to talk to him so he would understand them. I said good-bye to all of them and left.

Three weeks later, Fred's owners phoned me to say he was a completely different horse. He hadn't reared up at anyone and seemed to be much happier. Even his stiff leg didn't seem to be worrying him as much. They were thrilled and Fred was a relaxed happy and confident horse once again. He was going on outrides and enjoying himself.

I was happy to have been able to help.

# Chapter Ten

## SMURFIE'S REBIRTH AND GIFT

*Be proud of who you are.*
*Everything and everyone has a purpose.*
*We all need to be needed, we all want to be wanted.*
*Don't hide behind yourself and your insecurities.*
*Come out and share your love.*
*There are lots of souls out there that need it.*
*Be proud of who you are.*

**Lucy**

*We all have something to say.*
*You need to allow us to speak and*
*listen to what we have to say.*
*You might be pleasantly surprised.*

**Ballinore**

# SMURFIE'S REBIRTH AND GIFT

Shortly after the loss of my little Smurfie, I sent her photograph off to Amelia Kinkade to do a reading for me. She told me Smurfie would be back in this lifetime. Just as the psychic, whom I had gone to see two weeks previously, had told me.

I was very excited, and after that with every dog I saw, I wondered, "Was it Smurfie?"

How would I know? What if I missed her? What if I got the wrong dog? What should I do?

Years passed, then one day I picked up the "Animal Talk" magazine. The article on the cover was all about Siberian Huskies. I felt an urgency to go and see them. I wasn't planning to get a puppy, but I felt I just had go and see them.

When I got to the breeder, there were puppies running everywhere. They were all very cute. While I was sitting in the lounge enjoying the puppies, the owner brought in some other little puppies that were only ten days old. I looked at them, when suddenly, it was like a bolt of lightening hitting me in the chest. There she was, Smurfie, I just knew it was her. All my old feelings for her came flooding back. I picked her up and held her tight. She licked my face all over. My little dog had come back to me.

I bought her on the spot and visited her three times a week until the day came when I could take her home. I arrived that sunny day, picked up my little dog, which I had now named Candy, and drove home with her sitting on my lap.

When we got home, I took her inside and she already knew exactly where she was. She knew the house and garden, after all she had been here before, and had come home. All the other dogs got on well with her.

# SMURFIE'S REBIRTH AND GIFT

However, by the end of the first day, I noticed Candy had diarrhea, so I decided to take her for a check up. My vet at that time treated her for a tummy bug and sent her home. The next day I took her back, still the same problem, more antibiotics and home we came. After the fourth day with no success, I decided to get a second opinion and went to my present vet, who diagnosed Parvo virus and put her straight onto a drip. I was distraught.

My sister, Linda, and I went to visit Candy in hospital that night. Candy looked up at me and we connected on a deep level. I saw straight into her soul and she mine. We looked at each other without uttering a word for what seemed like an eternity.

The next morning, I arrived at the surgery to visit her. The vet came out to meet me, only to tell me that Candy had passed away minutes ago. I was devastated. I picked her up and took her home to bury her in my cemetery at home.

I have no memory of how I got home that day. I felt ill. Why had Smurfie come back just to go again after only a few weeks? I thought life was cruel. What had I done to deserve this? How was I going to go on?

Then one day, out of the blue, when I was least expecting it, came a very clear message from Smurfie (Candy). She said there was another dog out there that needed my love, and would give me great enjoyment. I should stop waiting for her. This is what she had come back to tell me. That same day I noticed an advert about puppies for sale. I felt drawn to them and so, that night, Linda and I went off to Kempton Park to see the puppies. The puppies were Siberian Huskies.

One of them came running up to me to play. He was beautiful, dark gray and white, with a dot on the tip of his nose. Riff-Raff (Raphael) had arrived, my gift from Smurfie. Linda drove home

# SMURFIE'S REBIRTH AND GIFT

while I sat with Riff-Raff on my lap, where I could admire him all the way home.

My little Smurfie had given me the greatest gift anyone could get. Although I still miss her terribly, I know she is with me, and I thank her for her sacrifice and her gift.

I love you, Smurfie, you are the light in the bottom of my soul.

I am honored to have shared your life with you.

Until we meet again, you will always be my shining star.

# MEDITATIONS & EXERCISES

*The secret to meditation*

*The secret to meditation is to relax and not make it an issue. If your mind is full of thoughts and you can't relax or still your mind, don't try too hard. Allow your thoughts to come (without trying to push them away), acknowledge them, and let them go. See them in a bubble of white light, or on a cloud, then watch them drift away.*

*It is better not to try to meditate at the end of the day when you are tired and your mind is full of everything you've done all day. Your mind is still rushing from the day's activities.*

*Early in the morning, when it is quiet, and before you start the day, is a good time. This will also give you added energy for the day's activities and will make it easier to still your mind.*

# MEDITATION
## Starting to meditate

Take your phone off the hook.

Sit in a quiet place where you won't be disturbed.

Take a few deep breaths.
As you exhale feel yourself slowly relax.
Focus on the end of your nose,
while continuing to concentrate on your breathing.

Do this for about ten minutes.
Then become aware of your whole body.

Stand up and stretch.

Keep doing this meditation exercise
until you find it easier to still your mind.

# MEDITATION
## 🐾 A relaxation meditation 🐾

Lie on your back with your eyes closed, either in the garden
where you can listen to the sounds of nature,
or in your house with soft music playing.
Whichever you prefer.

While concentrating on your breathing, start to relax your feet.
Move up, relaxing your calf muscles.
Relax your knees, your thighs, your hips.
Consciously relax your abdomen and pelvis.
Your solar plexus and chest.
Relax your neck and shoulders.
All the muscles on your face. Your jaw.
Let go of all the tension in the crown of your head and down
the back of your skull.
Feel yourself relax into your shoulder blades and all the way
down your spine into the small of your back.
Relax your buttocks.
Become aware of your skin, as it relaxes.
Take a deep breath and, as you exhale, let go of all the
remaining tension in your body.
Lie like this for about fifteen minutes, enjoying a state of
relaxation.

Slowly start becoming aware of the ground beneath your body.
The smells around you, the sounds you hear.
Open your eyes and look around you.
Stretch your arms above your head.
Stretch from the tips of your toes to the tips of your fingers.

Now slowly stand up.

# MEDITATION
## 🐾 Getting in tune with your inner self 🐾

Sit or lie down in a place
where you feel totally safe and won't be disturbed.

Close your eyes.
Picture yourself in a bubble of white light.
The white light is keeping you safe
and keeping all the outside noises out
(almost like a soundproof bubble).
All you can hear is the sound of your breathing
and the beating of your heart.
When you are totally relaxed
and focused on your inner self,
invite your animal guide to come on this journey with you.

Become aware of an animal approaching you.
This is your animal guide.
You walk out to meet him or her.

Picture yourself walking with your guide
through a field of lush green grass.
All around you are flowers of different colours.
There are butterflies, all fluttering about,
and birds soaring up above.
There is a slight breeze brushing
against your face as you walk along.

In the distance, you see a stream with a willow tree on its
bank. You walk up to this stream and sit down
under the willow tree with your guide
to listen to the sound of the water trickling past.
Absorb all the sounds and smells around you.

*continued...*

You are at peace. Sitting on the bank of the stream, you and
your guide look into the water at your reflections.
You drop a pebble into the water
and watch the ripples roll away.

Now put your hand into the water
and feel the cool cleansing effect of the water on your skin.

Sit back under the tree for a few moments
and just enjoy being with your animal guide
in this peaceful place.

When you have sat there long enough,
thank your guide for coming on this journey with you.

Lie on your back under the tree and close your eyes.
After a few moments,
start becoming aware of the floor beneath you.

Listen to the sounds around you.

Start to focus on your breathing
and the sound of your heart beating.

Become aware of all your surroundings.

When you are ready, open your eyes.

# EXERCISE 6
## Seeing colors to improve your telepathy

Sit opposite a friend or family member.

Close your eyes.

Take some deep breaths to relax.

Think of a color.

See if your partner can pick up the color you are thinking about.

Your partner might see a color, get a thought or even just a feeling.

Your partner should take note of the first thing that comes to his or her mind.

Now swap and let your partner think of a colour.

If you get it wrong, it doesn't matter.

You've got to start somewhere.

After a few tries, you will see how accurate you will become.

# EXERCISE 7

## Interviewing an animal to improve your imagination

This is a fun exercise that helps to remove inhibitions and to stimulate your imagination so that your intuition can develop.

Sit opposite a friend.

Your friend is going to interview you as if you're on air at a radio station. Choose to become one of your pets (for example, you have a Labrador named Rufus and you choose to be him). You will give your friend your name and he/she will introduce you by saying, "Good morning, listeners, I'm in the studio and with me I have, Rufus, a beautiful Labrador."

She will then start to interview you as Rufus, asking a range of questions, for example:

1.  Where do you live?
2.  Are you happy there?
3.  What are the people like who you live with?
4.  What is your favourite activity?
5.  What sort of garden do you have?
6.  Is there anything you fear?
7.  Do you have any health problems that you know about?
8.  Do you have any messages for your people?
9.  What would you like to tell the listeners of this station?
10. Thank you for joining us in the studio today.

Answer all these questions as if you were Rufus.

You and your friend don't need to stick to these questions. You can ask any questions you like.

When you have done your interview, change places with your friend and you ask the questions. Your friend will then become one of his/her pets and answer your questions.

# EXERCISE 8

## 🐾 Communication with cats 🐾
### at feeding time

If your cats suddenly disappear at feeding time, try to visualize them coming for their meal – and enjoying it.

Don't call, just sit and think about them coming to eat.

See how long it takes for them to get your message.

Next step is to see them eating from the opposite bowl. Observe them to see if they swap bowls.

# EXERCISE 9

## 🐾 Communication with dogs 🐾
### in the garden

When your dog has been playing or just lying in the sun outside for a while, ask him to please come and show you his face for a minute, so that you can see he is okay.

Wait for a while to see if he comes to you.

If he doesn't come, then show him a picture of himself coming to you for a pat.

Give him time to get your message and react.

# EXERCISE 10
## 🐾 An exercise in Gestalt 🐾

Choose only one of your animal friends. A dog, a cat or even a horse.

Slowly picture yourself becoming this animal.

Feel what it must be like to be this animal.

What can you see?

What can you smell?

What can you hear?

Who is with you?

What does the ground feel like under you?

Is it grass or sand?

Now see yourself becoming you again.

Go and look for your animal friend to see where he or she is. Are they in the same place that you saw?

# EXERCISE 11
## 🐾 Sending a holographic message 🐾

Close your eyes.

Picture an orange.

Once this picture has formed in your mind, open your eyes.

Now, once again, close your eyes and picture this orange a few feet in front of you, on the ground.

When this has been achieved, move the orange around the room by firstly placing it on top of the television set. Then picture it in the corner of the room.

When you have finished having fun with the orange, find one of your animal friends. It can be a dog, a cat, a horse, or even a hamster.

Think of who their favorite person is and create a holographic image by picturing this favorite person and placing them a few feet in front of the animal you have chosen for this exercise.

See what the animal's reaction is.

Practise makes perfect.

# EXERCISE 12
## 🐾 Reading a photograph 🐾

Find a friend who is willing to work with you.

Each of you has one photograph of an animal that you have chosen for this exercise.

Write down five questions that you want answered from your animal. You know the answers to three of these questions, and you don't know the answer to the other two.

Now swap photographs together with the questions.

You will now be reading your friend's photograph.

Move away to a quiet place so you can focus on the reading and slowly work through the questions. Write down everything that you pick up, no matter how silly it may seem.

When you are both finished, go over all the questions and answers with each other to see how accurate you are.

The more you practise this, the better you will get at it, and the more accurate you will become.

It is better to practise with an animal that you don't know very well.

# EXERCISE 13
## 🐾 Getting a reaction from a wild animal 🐾

When you visit a zoo or wildlife resort, find an animal you would like to work with.

Either walk up to the fence of the enclosure or stay in your car at a distance.

Say this animal is an elephant, then picture exactly what it is you want the elephant to do. See him looking directly at you. Picture him walking closer to you and standing there, ready for you to communicate with him.

Wait patiently to see how long it takes for him to respond.

If he does respond, thank him and ask him a few questions, for example:

- What is your purpose here?
- Do you have a message for me?
- What would you like me to do for you?

Give him time to answer each question.

Write down everything that comes to your mind.

Thank him before moving on to the next animal.

# EXERCISE 14
## 🐾 Coming home after an evening out 🐾

While driving home from an evening out, silently connect to your dogs at home.

Tell them that you are on your way home and will be there in a few minutes (or however long it will take you to get home).

Ask them please to meet you at the gate when you get home.

See if they are at the gate when you get home.

# EXERCISE 15
## 🐾 Telepathic messages 🐾

Start practising to give all your animals non-verbal (telepathic) messages.

Send them telepathic messages that you want to feed them, go for a walk or simply want them to come and sit with you in the garden.

Do this all just by thought.

Give them a treat, even if they don't connect with you.

All the exercises you do with your animals must be fun. Keep it light and don't get stressed if it doesn't seem to work. Give it time, it will work eventually.

*A very special lioness sharing a moment with her cub*

# THE LION HEART Project

*Be ready for change,
go forward in strength with love,
for it is through your heart
you connect with everything*

**Sly, the Lion**

*All information on the Lion Heart Project
has been provided by
Gail Kleinschmidt,
Founder and Director of the Lion Heart Project cc.*

# THE LION HEART PROJECT

The Lion Heart Project serves to provide education and upliftment for all communities. These programs take place in the form of positive interaction between man and the lions.

Our primary focus is on both linear and non-linear education. Linear education pertains to facts, information and research currently being done with animals, and especially with lion conservation. Non-linear education pertains to providing therapeutic and life-enhancing empowerment programs for people of all ages and walks of life.

Research has shown that animals provide both psychological and physiological benefits to man.

The Lion Heart Project has found that this interactive process between man and animals is profoundly beneficial – particularly when working with the White Lions of Timbavati (South Africa's ecological heritage) and golden lions.

Our aim is to continue providing this unique opportunity and service to communities at large. We are currently running a variety of highly inspirational and educational programs that are supported largely by sponsorship and donations.

Programs that have been run so far have covered working with disabled children in collaboration with Eco Access; working with schools and corporations; with individuals seeking therapeutic support; with vets, animal healer and animal communicators; and with conservationists.

Apart from offering these courses, we are also doing research in positive human/lion interaction with the following:

> • disabled people, blind and partially-sighted people, deaf people, people with a mental or intellectual disability,

# THE LION HEART PROJECT

- people with a physical disability
- people who are debilitated by depression
- adults or children who have experienced some form of trauma or abuse
- telepathic communication with animals, and
- empowering children with the knowledge of who they are within the context of working with lions.

All interaction with lions is done with the utmost respect for the animals and takes place in a safe and controlled environment for both the animals and the humans.

Man holds the key to restoring our ecological balance. Yet the Lion Heart Project has realized that information alone is not enough to turn the tide of the human heart. Humans require intense inspiration in order to make a positive and active change in any field of life, including conservation.

To this end, we hope to provide an environment where humans and animals can interact positively and where many of the deeper answers regarding our quest to conserve and restore balance may be elicited from these inspiring experiences.

Any sponsorship, funding and donations will be used to subsidize those schools and communities that are unable to participate financially.

For further information, you may contact one of the Lion Heart Project facilitators:

Gail Kleinschmidt
Founder and Director of the Lion Heart Project cc
Email: gail.lionheart@gmail.com

# IMPORTANT INFORMATION
## Protect
### YOUR CATS AND DOGS
### FROM HARMFUL FOODS AND SUBSTANCES

*Many people, while loving their pets,
are not aware of the potential harmful effects of
certain foods and medications.
We tend to want to spoil our pets.
Before giving your pet a treat,
it would be a good idea to check with your vet
whether or not the treat you are planning to give is
potentially dangerous to your pet.*

# DANGEROUS FOODS AND SUBSTANCES

Apple seeds and the pits of cherries, peaches, pears and plums. Apricot pits contain cyanide, which is poisonous.

Artificial sweeteners (such as those found in diet products) can cause a sudden drop in blood sugar, resulting in depression, loss of co-ordination and seizures. Unless treatment is given quickly, the animal could die.

Avocados (fruit, pit and plant) are toxic to dogs. They can cause breathing difficulties, fluid accumulation in the lungs or abdomen, or pancreatitis.

Caffeine (from coffee, coffee grounds, tea or tea bags) stimulates the central nervous and cardiovascular systems, and can cause vomiting, restlessness, heart palpitations and even death within hours.

Chocolate can cause seizures, coma and death. The darker the chocolate, the more dangerous, but any chocolate can be fatal. 25 grams of chocolate can kill a 14 kilogram dog. The symptoms might not show up for a while, leading you to believe everything is fine. However, death can occur within 24 hours.

Coins made from the 1980s to today contain zinc, which can cause kidney failure and damage to red blood cells. A dog that consumes even one such coin can become quite sick or even die if the coin is not removed.

Cooked bones can splinter and tear a dog's oesophagus, stomach or intestines.

Dairy products are high in fat, which can cause pancreatitis, gas and diarrhoea. A small amount of non-fat, plain yogurt is usually safe.

# DANGEROUS FOODS AND SUBSTANCES

Too much fat or fried food can cause pancreatitis. This includes foods such as bacon and ham.

Grains should not be given in large quantities and should not make up a large part of your dog's diet. But rice is generally safe if given in small quantities.

Grapes and raisins can cause kidney failure in dogs. As little as a single serving of raisins can kill them. If a dog doesn't eat enough to kill him in one go, small amounts over a period can cause severe damage.

Some insects accumulate cardiac glycosicles in their bodies from the milk woods (Asclepiadaceae) on which they feed. An example of this is the "Danouis Chrysippus", or Monarch Butterfly. Eating these insects can lead to cardiac failure.

Macadamia nuts can cause weakness, muscle tremor and paralysis. These symptoms are usually tempory.

Nutmeg can cause tremors, seizures and death.

Onions destroy red blood cells and can cause anemia, weakness and breathing difficulties. Even small amounts can cause cumulative damage over time. This includes onions or chives, in raw, powdered, dehydrated or cooked form. Large amounts of garlic cause the same problems as onions.

Even a small dose of certain human pain medication over a period of time can cause stomach ulcers. Cats do not have the enzyme to break down paracetamol, so any medications containing paracetamol are enough to kill a cat. It is vital to consult your vet before giving your pets any human medication.

Peach pits contain cyanide, which is poisonous.

# DANGEROUS FOODS AND SUBSTANCES

Raw egg whites contain a protein called avidin, which can deplete your dog of biotin, one of the B vitamins. Biotin is essential to your dog's growth and to the health of his coat. A lack of biotin can cause hair loss, weakness, growth retardation, or skeletal deformity.

Raw liver or too much cooked liver (three servings a week) can lead to vitamin A toxicity. This can cause deformed bones, excessive bone growth on the elbows and spine, weight loss and anorexia. Check the label on your dog food to make sure that it does not contain liver if you are already feeding liver.

Salty foods may make dogs drink too much water. This could lead to the development of bloat, which is fatal unless urgently treated. Bloat is where the stomach fills up with gas and, within several hours, it may twist and cause death.

Tomatoes can cause tremors and heart arrhythmia. Tomato plants are the most toxic. All parts of the tomato plant, except the tomato itself, are also toxic to humans.

Dogs cannot digest most vegetables (carrots, green beans, lettuce, potatoes or yams), whole or in large pieces. Potato peels and green potatoes are dangerous.

Walnuts are also poisonous to dogs.

Wild mushrooms can cause abdominal pain, salivation, liver damage, kidney damage, vomiting, diarrhoea, convulsions, coma or death.

# ANIMAL Organisations

*The following organisations depend on the involvement of volunteers and contributors.*

# ANIMAL ORGANISATIONS

## GAUTENG PROVINCE, SOUTH AFRICA

Animal Anti-Cruelty League
Tel (011) 435-0672 / (012) 362-4032

Bush Babies and other primates
Dr Martie Khun
Cell (082) 895-3555

Community Led Animal Welfare (CLAW)
Tel (011) 763-1638

De Wildt Cheetah and Wildlife Trust
Tel (012) 504-1921

Endangered Wildlife Trust
Tel (011) 486-1102

FreeMe Wildlife Rehabilitation Centre, Sandton
Tel (011) 807-6993
Cell (083) 558-5658

Mammals, reptiles, garden birds, etc
Friends Of The Cat
Tel (011) 442-7219

HAIG (Human Animal Interaction Group)
Eugenie Chopin
Tel (011) 884-3156

Hedgehogs
Dr Jill Drake
Cell (083) 601-0437

Highveld Horse Care Unit, Meyerton
Tel (016) 362-3587

# ANIMAL ORGANISATIONS

Irwin Animal Rescue
Tel (016) 590-1255
Cell (082) 950-4368
Plot 25 7th Street, Walkers Fruit Farms

Kitty Haven
Tel (011) 447-5275

Nature Conservation Department
Tel (011) 355-1459/1900

NSPCA
Tel (011) 907-3590

People's Dispensary for Sick Animals
Mofolo
Tel (011) 984-4340

Raptor Rescue, Johannesburg
Lianda Naude
Cell (072) 197-8134
Bird of Prey Rescue and Rehabilitation - Owl, eagle, falcon, hawk, buzzard, harrier, kite, hobby, kestrel, vulture, secretary bird, etc

SA Guide Dogs Association
Tel (011) 705-3512

SA National Bird of Prey Centre, Inanda
Tel (011) 648-3491
Cell (083) 585-9540

Society for Animals In Distress
Vorna Valley
Tel (011) 466-0261
Cell (082) 952-497

# ANIMAL ORGANISATIONS

Wildcare Africa, Pretoria
Chris Pears
Cell (082) 352-2579

World Primate Sanctuary
Cell: (082) 754-0998


## 🐾 KWAZULU NATAL PROVINCE, SOUTH AFRICA

Animal Anti Cruelty League
Tel (031) 209-7697

Animal Protection & Environmental Sanctuary
Cell (072) 306-5664
Including Primate Rehabilitation

Centre for the Rehabilitation of Wildlife, Durban
Tel (031) 462-1127

FreeMe Wildlife Rehab Centre, Howick
Cell (083) 378-3153

Nature Conservation Department
Tel (033) 845-1324


## 🐾 LIMPOPO PROVINCE, SOUTH AFRICA

Center for Animal Rehabilitation and Education, Phalaborwa
Irene McKenzie-Frazer
Tel (011) 468-3553 (Administration)
Cell (083) 659-3727

# ANIMAL ORGANISATIONS

Nature Conservation Department
Tel (015) 295-9300

Sanwild Wildlife Trust, Letsitele
Tel (015) 318-7900
Cell (083) 310-3882


## MPUMALANGA PROVINCE, SOUTH AFRICA

Nature Conservation Department
Tel (013) 759-5329


## NORTHERN CAPE PROVINCE, SOUTH AFRICA

Kalahari Raptor Centre
Kathu
Tel (053) 712-3576

Nature Conservation Department
Tel (053) 832-2143


## NORTH WEST PROVINCE, SOUTH AFRICA

Nature Conservation Department
Tel (018) 389-5048

# ANIMAL ORGANISATIONS

### 🐾 WESTERN PROVINCE, SOUTH AFRICA

Animal Anti Cruelty League
Tel (021) 700-6500

Animal Rescue
Karen De Klerk
Tel (021) 396-5511

Animal Welfare Society
Tel (021) 691-3702

Cart Horse Protection Association, Eppindust
Tel (021) 535-3435

CROW Joshua Baboon Rehabilitation Centre, Barrydale
Tel (028) 572-1643

Domestic Animal Rescue Group, Hout Bay
Tel (021) 790-0383

Monkeyland Primate Sanctuary, Plettenberg Bay
Tel (044) 534-8906

Nature Conservation Department
Tel (021) 483-3539

SANCOB, Cape Town
Tel (021) 557-6155

SPCA, Grassy Park
Tel (021) 700-4140

# ANIMAL ORGANISATIONS

The Emma Animal Rescue Society
Tel (021) 785-4482

Wildlife Rehabilitation Centre, Cape Town
Steph Wolf
Tel (021) 988-2502

WWF-SA, Stellenbosch
Tel (021) 888-2800

# ARTICLES WRITTEN ABOUT AND BY
## *Jenny Shone*

| | |
|---|---|
| *Free Spirit* | TV Program - July/Aug 2003 |
| *Smallholder* | Dec Jan 2004- 2005 |
| *Citizen* | 18 March 2004 |
| *Farmers Weekly* | Sep 2004 |
| *Odyssey E-Zine* | Dec 2004 |
| *Renaissance* | Oct 2004 |
| *Vaal Weekblad* | 11 - 13 Feb 2004 |
| *Natal Witness* | March 2005 |
| *Your Family* | April 2005 |
| *Vaal Star* | May 2005 |
| *Rooi Rose* | June 2005 |
| *Fusion* | Nov/Dec 2005 |

# RECOMMENDED Reading

Andrews, Ted. *"The Animal-Speak"*, Llewellyn Publishing, 2002.

Eason, Cassandra. *"The Psychic Powers of Animals, How to communicate with your pet"* Judy Piatkus Publishers, 2003.

Kinkade, Amelia. *"Straight From The Horse's Mouth, How to talk to animals and hear them talk back"* Crown Publications, 2001.

Roberts, Monty. *"The Man Who Listens to Horses"* New York Balatine Books, 1996.

Smith, Mark. *"Auras: See Them In Only 60 Seconds"* Llewellyn Publications, 1997.

Smith, Penelope. *"Animal Talk"* Council Oak Books, 2004.

Smith, Penelope. *"When Animals Speak"* Beyond Worlds Publishing Inc, 1999.

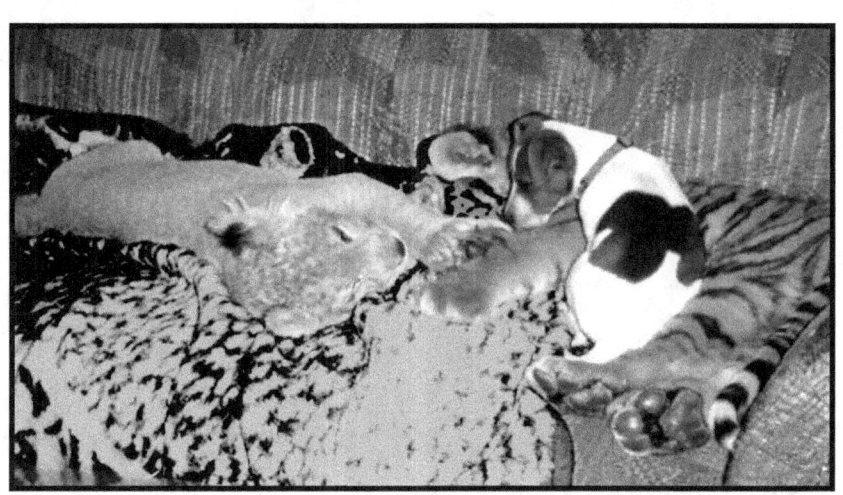

*The power of animal communication*

# Index

Abandonment
    what animals feel .................................................................. 167
Amputation ................................................................................. 49-50
Animal
    control ........................................................................................ 31
    feelings ........................................................................... 25, 29, 35
    individuality .............................................................................. 30
    pictures ................................................................................ 25, 26
    thoughts ................................................................ 22, 26, 97, 131
Anxiety
    reasons for ...................................................................... 29, 54, 81
Aura
    animals ............................................................................... 133-134
    energy field ........................................................................ 25, 98

Body Language ............................................................................... 27
Body Scan .................................................................................... 64-65

Cats
    de-clawed ................................................................................ 161
    in tune with self ......................................................................... 45
    learn from ........................................................................... 39, 47
    scratching furniture ................................................................ 161
Chakras
    energy centers within the body .............................................. 159
Clairaudience
    hearing ....................................................................................... 26
Clairvoyance

# INDEX

Clairvoyance
  seeing .................................................. 25
Clairsentience
  feeling ................................................. 26
Commands
  negative ............................................. 133
  positive .............................................. 133
Communication
  animal ................................................. 54
  jumping up ........................................... 2
  misinterpret ........................................ 40
Connecting
  animals ........................................... 56, 92
  animal guides ................................. 72, 73
  other side ........................................... 96
  photograph ........................................ 99
  soul .................................................... 99
  stranger ............................................ 139
Cruelty
  declawing ......................................... 161
  debarked ............................................ 31

Dangerous foods (see harmful foods)
Death
  concept .............................................. 99
  mourning ........................................... 98
  prey ................................................. 144
Disease
  auras ................................................ 159
  emotions .......................................... 159
  Gestalt ............................................... 55
  healing ............................................. 159
  imbalance ........................................ 159
  stress ................................................. 28

# INDEX

Eating Animals
    honor .................................................................................... 144
    respect ............................................................................ 144,145
Exercises
    An Exercise in Gestalt ............................................................ 187
    Coming home after an evening out ....................................... 191
    Communicating ...................................................................... 128
    Communication with cats at feeding time ............................. 186
    Communication with dogs in the garden ............................. 186
    Feeling emotions ...................................................................... 93
    Getting a reaction from a wild animal ................................... 190
    Getting permission ................................................................. 109
    Interviewing an animal .......................................................... 185
    Opening your heart .................................................................. 77
    Quieting your mind ................................................................. 52
    Reading a photograph ........................................................... 189
    Seeing colours to improve your telepathy ............................. 184
    Sending a holographic message ............................................ 188
    Telepathic messages .............................................................. 191

Fears
    abilities .................................................................................... 29
    animal fears ..................................................... 142, 161, 166
    of animals ............................................................................. 151
    reason ...................................................................................... 25
Feelings
    acknowledging ...................................................................... 148
    gut ................................................................................. 25, 167
Food
    dangerous/harmful .......................................................... 197-199

# I N D E X

Gestalt
    body scan ................................................................ 64
    Ginger ..................................................................... 82
    medical ........................................................ 55-56, 64
    method of ............................................................... 55
God Bless Our Pets ..................................................... 116

Harmful foods
    apple seeds ........................................................... 197
    apricot pits ........................................................... 197
    artificial sweeteners .............................................. 197
    avocados ............................................................... 197
    caffeine ................................................................. 197
    chocolate .............................................................. 197
    coins ..................................................................... 197
    cooked bones ....................................................... 197
    dairy products ...................................................... 197
    fat ......................................................................... 197
    fried food ............................................................. 198
    grains .................................................................... 198
    grapes/raisins ....................................................... 198
    insects ................................................................... 198
    macadamia nuts ................................................... 198
    nutmeg ................................................................. 198
    onions/garlic/chives ............................................. 198
    paracetamol ......................................................... 198
    peach pits ............................................................. 198
    raw egg whites ..................................................... 199
    raw/cooked liver .................................................. 199
    salty foods ........................................................... 199
    tomatoes .............................................................. 199
    vegetables ............................................................ 199
    walnuts ................................................................. 199

# I N D E X

wild mushrooms ................................................................ 199
Healing
    emotional ........................................................ 28, 159, 161, 169
    stressful ..................................................................... 158-159

Learning
    communicating ................................................................. 22
    hear ................................................................................ 26
    practice .......................................................................... 84
Listening
    depth ........................................................................... 134
    importance of ................................................................. 27
Lost
    Gestalt .......................................................................... 79
    remote viewing ................................................... 55, 56, 81
    techniques when tracking ............................................. 81
    tracking ..................................................................... 81-84
Love
    through the heart ............................................................ 46
    respect and animals ....................................................... 27
Lion Heart Project
    Facilitators ............................................................. 194-195
    Information ................................................................. 193

Meditations
    A relaxation exercise .............................................. 179-180
    Connecting to pets on the other side ........................ 105
    Connecting to the earth ............................................ 19-20
    Getting in tune with your inner self ......................... 182
    Meeting your animal guides ....................................... 73

# INDEX

Removing the cloak of negativity ............................................. 51
Starting to mediate ................................................................. 180
Messages (from the animals)
    Archie ................................................................. 44, 45,79
    Ballinore .............................................................. 168, 175
    Daisy ................................................................... 16, 39-40
    Ginger ........................................................................... 111
    Gizmo ........................................................................ 96-97
    Lucy ...................................................................... 170- 175
    Mona Lisa ................................................................. 33-34
    Penny ...................................................... 163, 166 - 167
    Rebecca ............................................................ 95, 164 -165
    Red ....................................................... 157, 165- 166
    Riff-Raff ............................................................... 21, 36-38
    Savannah ................................................................. 49-50
    Smokey ................................................................. 104-105
    Snoopy .................................................................. 38-39, 53
    Stacey ................................................................. 45-46, 53
    The Alpacas ..................................................................... 44
    The Cheetah .................................................................. 145
    The Lions ....................................................... 128, 135-137
    The Shark ...................................................................... 151
    Tootsie ................................................................... 126,169
Messages
    misinterpretation ..................................................... 40-41
Mind
    relax ............................................................................... 28
Mirroring
    actions ......................................................................... 142

Negativity
    aura .............................................................................. 158
    effects ........................................................................... 133

# I N D E X

Other Side
    heaven ................................................................................... 54
    mourning .......................................................................... 98-99
    soul .............................................................................. 111-112

Pain
    dealing with ......................................................................... 17
    determine .............................................................................. 55
    healing ............................................................................ 54-55
    health ........................................................................ 60, 156, 158
    undertstanding ...................................................................... 151
Pet Food
    health
Photographs
    communicate ........................................................................ 105
    Gestalt ................................................................................ 65
    photos on page ................................................................ 119-126
    readings ......................................................................... 99, 189
Pictures
    clairvoyance ......................................................................... 25
    communicate .......................................................................... 22
    Gestalt ................................................................................ 78
Prayer
    respect .............................................................................. 144

Rainbow Bridge ........................................................................ 109
Reincarnation
    rebirth .............................................................................. 111
Rescue worker ........................................................................ 105

# I N D E X

Silence
   mind .................................................................... 39, 46
Soul
   death .................................................................. 111, 143
       experience ............................................ 96-98, 111,112
   heaven ..................................................................... 54
   purpose ................................................................... 112
   connecting to ............................................................. 96
Spiritual
   beings ..................................................................... 44
   growth ................................................................... 114
Stories
   A healing story .................................................... 172-174
   Alan and his Rhino ordeal ..................................... 148
   Bella and the Workshop .......................................... 42
   Bird on a broom .................................................... 152
   Bird with a headache ....................................... 133-134
   Christopher Robin ................................................... 99
   Connecting to Whales ....................................... 151-152
   Dad and the Kudu ........................................... 128-129
   Diamond ............................................................... 140
   Dogs and the classical music .................................. 41
   Frodo ..................................................................... 48
   Ginger and Gestalt .................................................. 85
   Gizmo .................................................................... 96
   Horses at the fence ............................................... 164
   Horse with an abscess ............................................ 56
   Hungry snails ....................................................... 155
   Ian's story with the lioness and her four cubs ....... 145-147
   Jade ..................................................................... 141
   Keli and Lynette ..................................................... 62
   Kerry with hepatitis .............................................. 162
   London Zoo .......................................................... 130

# I N D E X

| | |
|---|---|
| Maggie | 102 |
| Meeting the Souls of all My Animals | 96 |
| Mishka's job | 28 |
| Mona-Lisa | 33 |
| My Spiritual Wolf | 57 |
| Nandi | 102 |
| Orange | 86 |
| Our Lion Park visit in March 2005 | 148 |
| Penny's book | 166 |
| Riff-Raff | 32-33, 36-38, 40 |
| Savannah | 49 |
| Smokey | 104 |
| Smurfie's story | 16-18 |
| Tales from a donkey | 170 |
| Talking to a cobra | 153 |
| Talking to Fly's | 154 |
| Tess | 100 |
| The Bees | 61 |
| The crested barbets | 129 |
| The day we met the Elephants | 137 |
| The Johannesburg Zoo | 147 |
| The Young Zebra | 138 |
| Tina and Charlie | 22 |
| Topaz | 103 |
| Tracking a lost kitty | 81 |
| Tracking Shingi | 83 |
| U Shaka | 150 |
| Venus | 139 |
| Victoria | 154-160 |
| Visiting the cheetahs | 150 |
| Vista | 41 |

# INDEX

Telepathy
  what is .................................................................................. 24
Termites
  removal .............................................................................. 149
The Dream ........................................................................... 137
Tracking (see Gestalt)
Trust
  Animal ................................................................................. 27
  Gestalt ................................................................................. 55
  Messages ............................................................................ 56

🐾

Visualisation (see Pictures)

🐾

NOTES

ALSO AVAILABLE FROM THE
## *Author*

*Meditation CD
and
Animal
Inspirational
Message
Cards*

Jenny Shone
Animal Healing Centre
P O Box 464 Walkerville 1876
E-mail: jenny@animalhealing.co.za
www.animalhealing.co.za

*(Prices available on request)*